FRACTALS

Applications in Biological Signalling and Image Processing

FRACTALS

Applications in Biological Signalling and Image Processing

DINESH K. KUMAR
RMIT University, Melbourne
VIC, Australia

SRIDHAR P. ARJUNAN
RMIT University, Melbourne
VIC, Australia

and

BEHZAD ALIAHMAD
RMIT University, Melbourne
VIC, Australia

CRC Press
Taylor & Francis Group
Boca Raton London New York

CRC Press is an imprint of the
Taylor & Francis Group, an **informa** business
A SCIENCE PUBLISHERS BOOK

CRC Press
Taylor & Francis Group
6000 Broken Sound Parkway NW, Suite 300
Boca Raton, FL 33487-2742

Printed on acid-free paper
Version Date: 20160826

International Standard Book Number-13: 978-1-4987-4421-8 (Hardback)

Visit the Taylor & Francis Web site at
http://www.taylorandfrancis.com

and the CRC Press Web site at
http://www.crcpress.com

Printed and bound in the United States of America by Publishers Graphics,
LLC on sustainably sourced paper.

Preface

It has been well established that healthy and stable natural systems are chaotic in nature. For example heart-rate variability and not heart-rate, is an important indicator of the healthy heart of the person. While there may be large differences in the resting heart-rate of two healthy individuals, it is important that this is not remaining monotonous but has significant variability. Over the past four decades, numerous formulas have been developed to measure and quantify such variability. This variability is often referred to as the complexity of the parameters and explained using Chaos Theory.

There are thousands of scientific publications on the application of Chaos Theory for the analysis of biomedical signals and images. We have attended many conferences and meetings where the relationship between the fractal dimension (FD) of biomedical signals and images with disease conditions, have been discussed. Many authors have demonstrated that there is change in the values of FD with factors such as age and health. The aim of this book is not to capture the details of these publications; because we are certain that the readers can access those papers directly and without our help. In our current world of information overload, we do not see the purpose for writing any book to be repeating publications that are already available.

When reading the numerous publications on the topic, one common shortcoming was observed; the authors gave numbers, formulas and in some cases, statistics. What they have missed out is the explanation to the concepts. The aim of this book is to provide the conceptual framework for fractal dimension of biomedical signals and images. We have begun by explaining the concepts of chaos, complexity and fractal properties of the signal in plain language and then discussed some examples to explain the concepts. We are aware that there are many more examples and research outcomes than are covered in this book. While we have attempted to discuss current research and examples, this book is not a replacement of your literature review on the topic.

We are hopeful that this book will help the reader understand the concepts and develop new applications. Once the fundamentals are understood, the human body could be recognised in terms of its chaotic properties. In such a situation, the measurements are not just numbers but quantification of the physical phenomena. We hope that this would be useful for engineers, physiologists, clinicians and lay persons.

Content

List of Figures

CHAPTER 1

Introduction

ABSTRACT

Biomedical signals and images have been found to be non-deterministic but describable as chaotic and having fractal properties. These are measured by obtaining the fractal dimension (FD) of the signal or image. The FD of biological signals has been associated with various health and age related factors. This chapter introduces the reader to the brief history of fractal analysis and examines this in terms of biological signals, images and data. The chapter sets the stage for detailed analysis in the subsequent chapters.

1.1 Introduction

Science attempts to model observations in terms of definitive laws and rules. It deals with supposedly predictable phenomena such as gravity, electricity, and biological processes. When these studies are undertaken, the system is simplified into a number of independent components, each described in deterministic terms. Such models are generally suitable for describing a large number of observations and most of our technology has evolved from such exercises. For example, the earth's surface was first thought to be flat; however, detailed analysis shows that the earth is round. Further analysis now demonstrates the relationship between the surface of earth and galaxies far away. While we all now know that earth is not flat, for many day to day applications, it is sufficient to model and explain most observations made by the naked eyes and for us human to perform many of our daily activities such as walking or driving. It also allows us to build our buildings and perform our other activities. However, it does not allow us to explore the Universe.

The three important laws of nature were discovered by Newton, though later were found to be inaccurate. Though these laws have been found to be inaccurate, they can still be used to explain most of the phenomena that are observed by us during our daily life and thus these laws cannot be considered to be incorrect. However, these laws are unable to provide the precision and clarity that may be necessary for certain purposes. One such example is the understanding of weather patterns. Seemingly similar conditions can lead to very different outcomes and deterministic computational models predictions appear to be very far from real observations.

Small causes can sometimes have abnormally large effects. This has been observed by philosophers, historians and scientists since time immemorial. Evolution in science is a result of continuous improvement in the experimental methodology and the ability to perform more exact measurements. Very often, the laws that describe the observations create methods and instruments, which on being used enable measurements to be performed more accurately, thereby negating the laws themselves. With the evolution of science, we know that the earth is not flat, and that there is an uncertainty in all measurements. However, our traditional mathematics that models the environment is designed to provide deterministic models, while many phenomena such as the weather and biological systems are not deterministic.

The term "chaos" had been used since antiquity to describe various forms of randomness, but in the late 1970s it became specifically tied to the phenomenon of sensitive dependence on initial conditions. Chaos is the science which explains when the outcomes do not appear to follow the natural laws leading to the unpredictable. It teaches us to expect the unexpected. The underpinning mathematics of Chaos theory allows the description of observations that appear to be unexplainable, even though the system seems to be well understood.

There are a number of reasons why many times the predicted outcomes appear to be very different from the observations. One reason is because the assumption of independence between different parts of the system is not accurate and there is a complex relationship between many seemingly independent elements. If we consider the body which is made up of a number of organs and each organ is made from individual cells. Each of these are independent, however they are also dependent on the rest of the body. For example, each cell requires the flow of blood, which requires a number of different organs. But these cells are also independent and many of these will continue to live after the body dies.

The second cause of large differences between the predictions and the observations are due to the variability in the initial conditions of the system. In most biological and natural systems, it is difficult to accurately identify the point in time that can be considered to be the starting point, thus determining the initial conditions accurately is impossible. While the definition of the start of life is given for the sake of legal or cultural reasons, it is near impossible to determine this from a scientific view point.

Chaos Theory describes nonlinearity and complexity of events and phenomena that are effectively impossible to predict or control, such as weather, or turbulence of a jet engine, or the states of the body. These phenomena are often described by fractal mathematics, which captures the infinite complexity of nature. Most natural systems and events exhibit fractal properties, including landscapes, clouds, trees, organs, rivers. Also many of the systems in which we live exhibit complex, chaotic behavior. Recognizing the chaotic, fractal nature of our world can give us new insight, power, and wisdom. For example, by understanding the complex, chaotic dynamics of the atmosphere, a balloon pilot can "steer" a balloon to a desired location. By understanding that our ecosystems, our social systems, and our economic systems are interconnected, we can hope to avoid actions which may end up being detrimental to our long-term well-being.

Biological systems and most natural systems can often be treated as systems within systems. Similar to the concept of seeing the Universe as a giant atom, biological systems can, to an extent, be treated as having scale dependence on the observer. These properties may be in terms of spatial, temporal or other dimensions and can be referred to by their self-affinity and are described by fractal geometry.

According to Edward Norton Lorenz [1], the entire universe is connected, and the movement of air due to the fluttering of the wings of a butterfly in one location can be the cause of a storm in a place that is very remote. Similarly, biological systems may appear to have separate organs, but the entire body is a single entity and a small change in one part of the body could lead to major changes in a different part of the body. Often, this cause and effect may not be evident when we look at an individual organ or part of the body. It is important that the entire system should be considered as a whole along with considering the individual parts for accurate diagnosis and predictions.

Biological systems are known to be unpredictable and it is often difficult to predict the outcome or response of the body to treatment or to a change in circumstances. Chaos theory has demonstrated that small difference in the initial condition can change the outcome of an experiment very significantly. As the initial conditions are difficult to know precisely, and it is often difficult to identify what should be considered as the starting point, the final outcome is very unpredictable in biological systems. Often, what appears to be disorder and random behavior is not because of lack of order but due to this unpredictability and interconnectivity.

While outside the scope of this book, there is the concept of the difference in the psychological response between different people. The fundamental laws that govern all people are the same, but the behavior of different people to the same situation can be very different. It has been found that even identical twins brought up under identical conditions can behave very differently. This is now understood in terms of chaos theory, which explains that seemingly similar initial conditions would have sufficient differences that leads to large differences, and makes the two people behave extremely differently.

Fractals are an ever continuing pattern, a pattern that is infinitely complex and is based on self-similarity, with the underlying process being simple. Fractals show a system that is seemingly based on very simple principles but leads to very complex structures. Fractals are suited to describe the chaotic behavior and are effective in describing systems such as biological systems. They have also been adapted to describe and develop music and art, to study natural objects, and have also been used in attempts at giving rigor to concepts such as beauty.

1.2 History of Fractal Analysis

The concepts of fractals and Chaos have been discussed by scientists and philosophers for a long period of time. However, it was the availability of computers, especially with graphical displays that allowed the formulation of the fractal concepts. Such displays have also provided the strength of imagery and based on these displays, the abstract concepts have become easier to understand for even lay people, and are now commonly accepted.

Fractal geometry, associated math and analysis techniques are largely attributed to Benoit Mandelbrot, a Polish born French mathematician (20 November 1924–14 October 2010). While the fundamental concepts had been discussed far earlier in physics, and perhaps these concepts have been discussed in Chinese and Indian philosophies a few thousand years ago, he was responsible for formulating the concept and providing the rigor, though in an unconventional style. He studied and demonstrated the scale invariant properties in nature. He created the associated concepts of self-similarity and later in 1975 coined the term, Fractals, and also associated this with the concept of roughness. The word, fractals, is from Latin and means, fractured, and implies the inherent complexity at changed scales. He demonstrated this concept using graphical displays of fractals which showed how visual complexity could be created from underlying simple rules.

1.3 Fundamentals of Fractals

A pattern, with the repetition embedded in it is called Fractal, and may be from a natural phenomenon or a numerical set. It is also known as expanding symmetry or evolving symmetry and has been shown to describe the power law. If the replication is exactly the same at every scale, it is called a self-similar pattern. There are number of computational examples that provide the visualisation of this phenomenon such as Menger Sponge and Koch Curve. However, in real life, the phenomena are not exactly self-similar, but nearly identical at different scales and are called Fractals.

Geometric figures such as a rectangle or any other polygon has the area change by the square of the factor by which the lengths of the sides were changed. Thus, if the length of the side of a square is doubled, its area increases by 4, or 2^n, where $n = 2$, the dimension of the polygon. However, when the length of the fractal is doubled, the area increases by a factor which is not an integer, or n is not an integer but a fraction and is the fractal dimension of the objects exhibiting fractal geometry.

Shapes or functions that describe fractal geometry (or data) are generally not differentiable. A line is generally considered to be 1 dimensional. However, a line corresponding to fractal dimension would have resemblance to a surface and would have a dimension greater than 1.

1.4 Definition of Fractal

There is no precise definition for fractals. Perhaps the best description for fractals is that these have (1) self-similarity, (2) iterated, and (3) fractal dimensions. These are not limited to geometric patterns or mathematical expressions, with many examples in nature that may describe functions of time.

Conceptually, fractals can be considered to be associated with repetitive procedures, and while the fundamental procedure is simple, the final output is very complex. A simple example of a fractal is based on an equilateral triangle which has each sides of 1 unit length. Further, on each edge of the triangle, if a new (smaller) equilateral triangle is included, sides of these triangles will be 1/3 units. Repeat this process to each of the new triangle sides, and this time the length of the sides of the new triangles will be 1/9 units. If this process is continued till infinitum, we now have the fractal. It can be observed from the Fig. 1.1 that magnification (or reduction) along any edge results in the same shape.

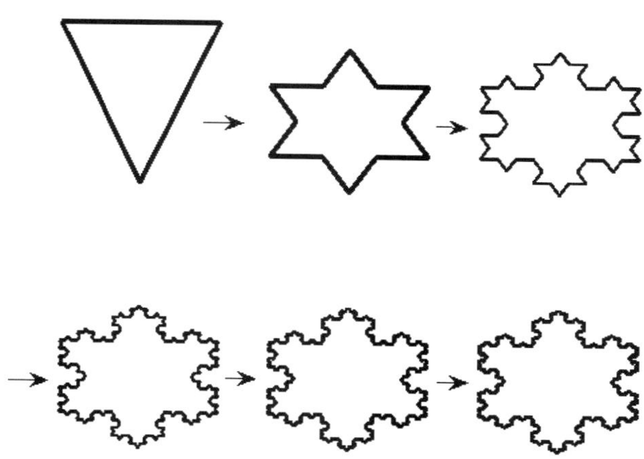

Figure 1.1. Fractal object with an equilateral triangle.

The area of a fractal shaped object or figure cannot be computed using concepts of polygon geometry, but needs to be obtained by summing the area of the individual areas, in this case, the individual triangles. Based on the visualisation of this figure and geometric summation, the area of the figure will converge to a finite number. However, the length of the line, or perimeter, could continue to increase as the number of iterations increases, and theoretically, this could increase to infinity.

Using geometric series, the length of each side would increase by a factor of 1/3 for each iteration, or the perimeter = $3^*(4/3)^m$, where m = number of iterations. It has been empirically found that the distributions of a wide variety of natural, biological, and many man-made phenomena approximately follow a three power law. This explains number of observations as when there is relatively large reduction in the size of some object such as the cross-section of a pipe, the reduction in throughput is relatively small. This is described by the relationship $A^{3/4}$ law and has been found to be associated with the self-similarity within itself.

1.5 Complexity of Biological Systems

No two bodies are identical and within the body, there is never exact symmetry. The body is nearly laterally symmetrical and the heart nearly beats regularly. However, the important word is; nearly. But within the short observations, and in controlled conditions, our physiology may be described as if deterministic and for most measurements the body does appear to be symmetrical. It is often the small differences that exist in different dimensions of the body or in different selections within the body that makes it robust, resilient, and perhaps beautiful.

Bodies are complex structures which can be described as fractal. Whether we see our skin surface, the neurons, the circulatory system, or our retinal vasculature, all of these follow a tree structure. These follow an approximate self-similarity and can be defined in terms of their fractal properties. One common outcome of such a structure is that it allows for packing very long or wide structures in a small volume. Thus, based on such branching tree structure, the lungs maximize the surface area, thereby enhancing our capabilities for exchange of gases, while enclosed in a small volume. The lungs have an area of a full tennis court while in a few cubic centimeters of space. Such enhanced area allows for greater exchange of gases, making breathing more efficient.

Another section of our body that packs a large number of connectivity in a small volume is the brain. One aspect of evolution is the high degree of interconnectivity between neurons in the brain, and the very large area of the cerebrum. The brain's geometry and inter-neural connectivity has been found to be fractal. Yet another example is the circulatory system, where with a natural tree structure and occupying only 3% of the body volume, the blood vessels reach all parts of the body.

Biological systems have been found to follow the ubiquitous Three-Quarter Power Law which is described by Fractal geometry. This particular power law is based on the cube of the fourth root. Many three-quarter laws have emerged from the measurement of seemingly unrelated systems, modeling the relationship of different structures.

It has been found empirically that the body's anatomy and physiology follow fractal principles. Looking at the anatomy, our cerebrum shape, neural connectivity, lungs, and capillaries, are describable by their fractal dimensions and tree structures. The physiology of the body also follows the same principle. While our heartbeats are often described as regular and rhythmical, these are fractal in nature. It is actually very important that our heartbeats are not exactly regular; the variation in the beat is responsible for reducing dramatic fatigue and wear and tear on the heart. Research has discovered that there is a reduction in the fractal properties of our heart beat with disease and with ageing. The same applies for all other parts of the body.

1.6 Fractal Dimension

Fractal dimension (FD), first conceptualized in 1975 by Mandelbrot [2], is an index for quantifying the fractal properties of an event or object. Often the terms such as fractal dimension, complexity and information are used interchangeably.

1.7 Summary of this Book

After this introductory chapter, the second chapter briefly describes the fractal dimensions concepts, and some of the algorithms that are used for computing FD. The subsequent chapters are divided in two major sections; the first of which describes the fractal dimension of the physiological parameters while the second section is devoted to the fractal dimension of the anatomical measurements. Finally, case studies are provided to describe some of the recent research outcomes that show the healthcare relevance of using fractal analysis.

References

1. Lorenz, E.N. 1963. Deterministic nonperiodic flow. Journal of the Atmospheric Sciences, 20(2): 130–141.
2. Mandelbrot, B.B. and J.A. Wheeler. 1983. The fractal geometry of nature. American Journal of Physics, 51(3): 286–287.

CHAPTER 2

Physiology, Anatomy and Fractal Properties

ABSTRACT

This chapter discusses the concepts of the relationship between chaos theory, complexity and the biological systems. It first shows why the traditional mathematical concepts that are based on Calculus are highly limited for biomedical analysis. It then introduces the fundamental concepts of chaos, complexity and self-similarity. This chapter also describes the use of fractal geometry and the differences from the traditional calculus. The chapter introduces the methods of measuring and quantifying complexity and chaotic properties using entropy and fractal dimension. Some of the commonly used methods to measure fractal properties are described and their properties examined in relation to biomedical applications; biomedical imaging and biosignal analysis.

2.1 Introduction

Often the body is described to be laterally symmetrical, and our physiology is considered to be periodical in nature. However, the truth is different from this; our bodies are not symmetrical, nor are our physiological parameters periodic. Most people have a significant difference in the lengths of the right and left legs and arms, and our heart beat is not exactly periodic.

While the body may not be exactly symmetrical and the physiology may not be exactly periodic, these assumptions are in most cases suitable for describing observations. Our clothes are made symmetrical and most of us do not notice any difference between the lengths of our right and left leg. And when we visit a clinic, our heart rate is monitored over a short period of time and seems to be reasonably periodic. However, it is well known that the body is asymmetrical and thus the symmetry only serves the purpose of simplification. Our dominant side muscles are significantly stronger and larger than the other side. Similarly, our physiological parameters are not periodic, and research has demonstrated that when the parameters become very periodic, it is not sign of good health. More recently, research has measured the variability of the parameters and identified the relationship of such variability with healthy conditions. The lack of periodicity and the asymmetry are the basis of natural phenomena. Traditional mathematical concepts of Calculus are unsuitable for studying these.

There is yet another factor that is relevant for describing biological systems. We are not modular and compartmentalized, but a single unit. Actually, considering us as a single unit is also false, because we are connected with our environment. We now understand the high level of interdependence between different organisms. And, looking at any biological system shows close similarity and resemblance within the species, yet careful observations shows that each sample is unique. Calculus and related mathematics are unable to describe us in details and can best describe us in terms of some overall expectations.

Chaos theory, a mathematical concept, refers to the principles that examine such variability and small differences between seemingly identical objects. It overcomes some of the limitations of Calculus, and explains how this lack of symmetry underpins the otherwise branching order, and why counter-intuitively a chaotic system makes the system stable. While each organism is a very complex structure, it can be viewed as a combination or network of simpler structures, and each of these structures being very similar. Just like no two leaves on a tree are ever identical but very similar, similarly no two blood vessels are identical. The combination of these results in a complex branching system can be modeled approximately by its self-similarity.

One common observation of biological systems is that while there is an underlying similarity, there are vast differences between two samples. This is observed within a single organism or between two organisms of a species. Thus, while the body is made of similar cells, and all cells are very similar, the cells of the body create unique parts or organs of the body. Similarly, while all human bodies are very similar, each of us is unique and we all appear to be different. We have different sizes, gender, and color and so on. Calculus is unable to describe these differences.

Chaos explains both these concepts; the self-similarity within a system, and the cause of the large divergence in similar systems in similar conditions. It shows that when a simple system is replicated and interconnected many times, the resultant system appears very complex and can perform complex functions. The theory also shows that when there are two identical systems, small differences in the initial conditions can lead to large divergence and thus the final system is very different. These concepts are discussed below.

2.2 Conceptual Understanding

Modern medicine has provided in-depth solutions for many ailments and understanding of the functioning of individual organs. Modern medicine has developed large number of specializations, and super-specialization in very specific topics. Thus, there are clinicians who not only specialize in cardiology, but may be the specialists specifically in the analysis of the left ventricular disease. Such specializations are extremely useful and form the basis for the capability of our modern medicine to look after patients and provide longevity as well as quality of life. The miracle of modern medicine is often attributed to the development of super-specialists. However, what is lacking is the holistic approach to health of an individual, and the society as a whole. Chaos theory provides a means of looking at the complete picture, and its style is of making it inclusive rather than exclusive.

Chaos is often used very loosely in our spoken English, generally indicating something that is disorderly. While the mathematical approach is based to describe the disorder, it demonstrates a very stable system. It is a description of the stability and order in the seemingly disorderly system.

For a closer look at this, and to answer the question, '*What is chaos*', consider it as a way to describe the complex phenomena. It is the mathematical approach that illustrates that it is possible to get completely random results from normal simple functions or equations. It is the bridge of finding order in what appears to be completely random.

Many researchers have demonstrated that systems that can be described using chaos theory are stable systems. This has been observed in large social systems, in nature and in biological systems. Studies have also shown that music and images that are based on chaos theory are generally more appealing to lay people and physiological parameters that are chaotic generally represent better health compared with systems that are well defined.

To measure the chaotic nature, measures such as Fractals and Entropy have been developed. These are closely related measures, and there are large numbers of algorithms that have been developed to measure these. The concepts of chaos theory, complexity, fractal properties and entropy are broadly discussed in the next few sections. The following section examines the concepts that relates the fractal properties of natural events and objects and relates these with biological parameters.

2.3 Chaos, Complexity, Fractals and Entropy

Fractal properties, chaos, complexity and entropy are often used synonymously. However, these are obviously not the same, but do have commonalities. What do they mean, and what is the relationship between them? This section examines these questions, and explains the different ways of measuring these properties.

Calculus has provided an excellent method for modeling many observations, and can be considered to be the basis for most of our modern science; physical, biological or social. It has given the means for understanding range of concepts such as thermodynamics, electricity and electromagnetism and provides means for computerized analysis of speech, and discovering DNA. It even provides the basis for understanding the concepts of relativity and quantum physics. In terms of Calculus based sciences, all problems could be analyzed completely, even though the exact answer may not often be possible.

What Calculus has been unable to explain is that when an experiment is repeated, the outcomes are similar, and follow the same principles, but they are never exactly the same. Most times these differences are small and within the acceptable range of error, and often describable in terms of the statistical distribution of the inputs. Many processes such as a manufacturing system generate the outcomes that are nearly identical and can be usually considered to be the same. Consider a process manufacturing screws, where we would always expect to get identical products.

There are many other times when a process is repeated, but the outcomes are very different even though all the parameters appear to be the same. There are number of examples, both in science, math and social sciences. One popular example is that of identical twins, who, having had identical conditions during and at birth, grow up very different. There are unlimited examples of such behavior ranging from financial markets to vegetation and flight paths. Even in concepts of neural networks, the outcomes of well-planned software outcomes can diverge significantly. Calculus is unable to describe such differences and would describe such experimental outcomes in terms of outliers or erroneous. However, these happen very often and many natural phenomena and mathematical modeling lead to such outcomes.

Fractal geometry describes the irregularity or fragmented shape of natural features as well as other complex objects where Euclidean geometry fails. This phenomenon is often expressed by spatial or time-domain statistical scaling laws and is mainly characterized by the power-law behaviour of real-world physical systems.

2.4 Chaos Theory

Chaos is a purely mathematical concept that overcomes the limitations of Calculus in the representation of natural and biological systems. Calculus assumes the lines are either straight or curved and describable by linear or quadratic or polynomial equations, natural systems do not strictly follow these laws. Thus, predictions based on Calculus are often imprecise and sometimes in natural systems, the predicted values and the observed values may disagree vastly. Most natural edges have an associated roughness which may be observable only at finer scales, and thus different edges may have a level of similarity but are never identical.

Chaos can be defined in any dimension, though the most common representation is in time and space. An object is defined in space; and if it is chaotic in space, it is fractal, or fractured. Unlike calculus, its edge or surface cannot be defined by a polynomial equation, and has a roughness, however small be the resolution. Unlike calculus, it does not respond to the concept of going to the limit.

Chaos is the property of systems that have non-linearity and interdependence, where order appears to be similar to natural disorder. Such a system is highly sensitive to the initial conditions and a small change in them will lead to divergence, where the possible states are radically different from each other. It is generated by a dense network of very simple systems that repeats and evolves, such that it is a necessary property of systems with the potential of evolving and growth, such as biological systems. This also may be described in terms of 'fold and stretch' a phenomenon which obviously leads to self-similarity. This fold and stretch provides the ability of a system to evolve, a process often referred to as 'emergence'.

Chaos in time domain is similar to chaos in space domain. A dynamic system is one whose conditions change with time. If we consider a typical dynamic system defining an object moving in space over time, then we can generally identify the trajectory of the object if we know the state of the object and the equations that govern its movement over time. By this principle, we are declaring that we could know the state of the object forever. If this was valid, every well trained golfer would be able to predict the flight of the ball, and the movement of the ball after the trajectory was obtained would remain the same. And even if there are small differences in the swing of the golfer, there would be anticipated small differences in the resultant location where the ball comes to a halt. However, we know that this is not the case, and the ball could end up in very different locations. While Calculus will struggle to explain these differences, Chaos theory shows that small initial differences and small differences in the environmental conditions could lead to these large differences.

Time domain chaos or time-chaos is attributed to the outcomes being sensitive to initial conditions. This explains that the trajectory of an object (or an event) could alter vastly when there may be small difference in the initial conditions. Lorenz work showed that the outcome of the two could, though very similar and close to each other at the beginning may eventually diverge exponentially away from each other [1]. Such sensitivity to initial conditions demonstrates that definitive predictions offered by Calculus, often referred to as reductionism, are not suitable for many situations such as for predicting the biological or environmental outcomes. Small uncertainty that may exist in the initial conditions could grow very quickly with time and become so large that the measurements and predictions appear to be unrelated and the computation has no real relationship with the actual state of the system. Thus, even if the state of the system is known with precision at a given time, the future trajectory cannot be predicted well. Nature is full of examples of this, including the weather patterns and biological systems. Chaos is a means of describing systems that have repeated self-similarity with sensitivity to the initial conditions, and describes many natural and biological phenomena.

The difference between Calculus and Fractal geometry is because of the localization approach by Calculus. These two approaches would be similar and converge if the holistic approach is taken. Thus Calculus and Chaos can theoretically converge, if we can consider the complete Universe as a single entity and all information from the beginning. However, we are aware of the difficulty in such, and know that it is impossible for all the variables to be available. Where Calculus attempts to identify and perform predictions, chaos describes the possibility of divergence of the outcomes of repeated experiments.

2.5 Complex Systems

When we consider a complex biological system such as the body, while it is important to obtain the details of individual parameters, it is also essential to see the system as a whole. Such a system is a complex system, where seeing one aspect of the organism or organization does not give the picture of the complete object or of other aspects of this object. Thus, knowing about the leg does not convey any information regarding the head. Such systems are not homogenous and do not have similar properties at different locations even though when seen at finer scales they may appear similar. To understand such systems, it is essential to see the factors that contribute to the complexity of such a system.

There is no real definition of complexity or complex systems, however, there are some accepted factors that are commonly acknowledged as properties of complex systems. One important property is that Complex systems always have many constituents or parameters, and the interaction between them is nonlinear. Another important factor is that these are not independent but there is often a high degree of interdependence. However, many times, complex problems have been modeled using linear equations, where the different parameters are considered to be independent and the descriptors are considered to be linear. Such models may be useful in describing and listing the number of parameters but are unable to accurately predict the true outcomes. One such example is the population based studies to predict risk of diseases such as cardiovascular disease. The Framingham equation was developed using a large database developed from the region of Framingham in USA and this lists some of the major parameters that contribute towards the risk of disease [2]. However, the linearity in such an equation results in poor sensitivity and specificity of risk of disease [3,4]. Further, it is quite evident that the parameters are not independent but interrelated. For example, the BMI, gender and blood pressure are not independent but interdependent. This example demonstrates that while Calculus is suitable for identifying the health parameters that are important to be considered, the prediction or risk assessment may not be accurate because these parameters are interdependent and do not interact linearly in complex situations.

One common observation of complex systems is that number of such system can be approximated or simplified in terms of a complex network of smaller and simpler modules, and where the simpler and the complex have a similar appearance. Thus, complex systems appear to have similarities at different scales. However, such systems also exhibit an emergence behavior, where the focus appears to change at different scales, and going from finer to coarser scale highlights difference in its behavior.

2.6 Entropy

When a system evolves, its entropy appears to increase. This also means that the reduction of entropy is caused by the effect of external factors that are causing this to happen. The concept of entropy is used to describe the disordered behavior of the system and is often used in the context of thermodynamics, and referred to as the second law of thermodynamics. This is also used to describe the concept of information. In this context, it is defined as the amount of additional information that is required to describe the state of the system.

To study complex systems and identify a measure of complexity requires obtaining statistical analysis to measure the level of disorder. The level of disorder may be in many different aspects such as disorder over time, or within space, or other descriptors. It also could be a measure of disorder between different objects or of the same object.

Entropy is a scientific term that describes disorder, both within a system and between different systems. Entropy quantifies the disorder which is required to study and compare systems. There are number of different measures of entropy and while there are many differences, one common factor is that entropy of a system is always increasing as long as the system evolves. When the system reaches equilibrium and stops evolving, its entropy does not change. However, when the entropy of a system reduces, that indicates that there are external factors that are causing changes. Thus, while a system left isolated would have an increase in its disorder, and its entropy will increase, effects from external sources may cause the system to become more ordered and its entropy would reduce. Some of the important measures of entropy are mentioned below:

a) *Shannon Entropy*: It is a measure of the average information content when the value of the random variable is not known [5] and is denoted by the following expression [5,6]:

$$S(X) = -\log_b \sum_{i=1}^{m} p_i \ln p_i \tag{2.1}$$

where X is a discrete random variable, p_i is the probability of the event $\{X = x_i\}$, and b is the base of the logarithm.

b) *Rényi entropy*: The Rényi entropy [7] of order q is defined for q ≥ 1 (for q → 1 as a limit) by the following equation:

$$S_q(X) = \frac{1}{1-q} \log_b \sum_{i}^{m} p_i^q \tag{2.2}$$

where X is a discrete random variable, p_i is the probability of the event $\{X = x_i\}$, and b is the base of the logarithm.

This measure is a generalization of Shannon entropy and denoted as one of the families of functions for quantifying the diversity, uncertainty or randomness of a system [5,7].

c) The *Kolmogorov entropy* is an important measure which describes the degree of chaotic nature of the systems [5,8]. The generalized Kolmogorov entropy Kq can be defined in the space where it is divided into the n-dimensional hypercubes of side r (at time intervals Δt), by the following equation [5,8]:

$$K_q(X) = -\lim_{r\to 0} \lim_{\Delta t\to 0} \lim_{N\to\infty} \frac{1}{N\Delta t} \frac{1}{q-1} \ln \sum_{i_1,i_2,i_3\ldots,i_N}^{m(r)} p^q_{i_1,i_2,i_3\ldots,i_N} \tag{2.3}$$

where $\{X = x_i\}$ is the discrete random variable and $x_i = x(t = i\Delta t)$. $p_{i_1,i_2,i_3\ldots,i_N}$ is the joint probability that the x(t = Δt) is in the box i_1.......and x(t = NΔt) is in the box i_N.

2.7 Fractal and Fractal Dimension

Fractal refers to a phenomenon, an object or a signal, or a mathematical description that has a repeating pattern that has similar display at every scale. Another way of describing fractals is as an expanding or evolving symmetry. When the replication is identical at every scale, it is called a self-similar pattern, and there are number of mathematical examples such as Menger Sponge or Sierpinski triangles. In nature, however, Fractals are not exactly but nearly the same at different levels.

Fractal dimension can be considered as a measure of the fractal properties. This is often required to compare the fractal properties and thus provides the basis for quantification of these properties.

If we consider the question, *What is a dimension in the spatial domain*? In Euclidean space, a line is considered to have one dimension, while a rectangle has a dimension of two. This is because there is only one linearly independent direction in a straight line, and two linearly independent directions in a plane. In a line there is only one way to move while the plane has dimension of two because there are two independent directions. Similarly, cube has three dimensions: length, width, and height. An alternative way to view the concept of dimension is for a self-similar object. The dimension N is the exponent of the number of self-similar pieces created with magnification factor $\mathbf{N}$. Thus, when a square is divided in 3^2 segments, the magnification is three, and the dimension is two.

However, in Euclidean space, the above is no longer true when a line is curved, or a rectangle is in a curved plane. The dimension of a curved line in a plane is not one. However, it is also not two, because the curve does not give the freedom for moving in two directions. Fractal dimension tells us how complex a self-similar system is. It measures how many possibilities are permitted in the set.

A multifractal system is a generalization of a fractal system in which a single fractal dimension is not sufficient to describe the dynamics. This describes the fractal dimension of a subset of points of a function belonging to a group of points and thus an object or signal that has complexity at multiple dilation factors. This concept was first applied to problems of turbulence. It suggests that there is an order, even when there are irregular behaviours at irregular points [9].

2.8 Computing Fractal Dimension

Fractal dimension is the rate of change on the logarithmic scale of the measured quantity with respect to the resolution or scale. The direct method to compute the fractal dimension is referred to as the Hausdorff equation or Hausdorff- Besicovitch equation. This is based on the assumption that the change on the log scale is linear and continues at the same rate. However, this assumption is neither accurate, nor is it generally possible to compute directly and hence it is essential to estimate the dimension, *N*, approximately. There are many different ways of estimating the fractal dimensions. Largely, these can be performed in time domain or spatial domain, or by considering the Fourier transform of the data and performing the analysis in spectral domain.

Number of methods to estimate have been developed over the years, and each of these have some unique properties. Although, each method is different, they follow the same principles. For image analysis, the fundamentals of this can be described in five steps:

1. Identify the feature of the object being tested and the scale with which the measurements will be made.
2. Measure the quantity at number of scales, and observe the limits of any relationship.
3. Plot on the log scale the measured quantity and the scale.
4. Estimate the linear approximation of the relationship.
5. The slope of this line is the fractal dimension (FD) of the object or event.

These measurement techniques broadly fall under the following categories; box-counting method, direct area measurement method, spectral based measurement and Brownian motion method. Some of these are briefly described below and more methods are explained in Chapter 3.

2.8.1 Box-counting

There are number of algorithms that are considered as box-counting (BC) fractal dimension techniques. One common step is that a mesh is created on which the image (or signal) is superimposed. The underlying principles of this technique are simple and implementation is generally easy.

The Box counting technique was reported by Russel et al. in 1980 [10]. In the simplest form, the binary signal is covered with boxes of length *r*, and FD is estimated as:

$$FD_{BC} = -\lim_{r=0} (\log (N(r))/\log (r) \tag{2.4}$$

where *r* is the scale, and *N*(*r*) is the number of boxes required to cover the signal.

One major limitation of this approach is that it requires the signal to be binarized. There are other theoretical limitations in this approach as well. To overcome these shortcomings, the Differential box-counting method (DBCM) was developed in 1995 by Sarkar and Chaudhari [11]. One of the biggest advantages of DBCM is that it does not require the signal or image to be binarised. Instead, the signal is partitioned into boxes of various size r and $N(r)$ is based on the largest difference in the box. The difference is computed between the minimum and the maximum values of the grey levels in a given box and the process is then repeated for all boxes. One of the major limitations of this method is that the outcomes are very sensitive to the choice of the lower and upper limit of the scale. It is important to estimate the limit of the scale prior to the test. An incorrect estimate of the lower and upper bound of the scale can lead to highly erroneous outcomes.

2.8.2 Power spectrum fractal dimension

Power spectral fractal dimension, also called Fourier Fractal Dimension, is based on the properties of Brownian motion, commonly observed in the 'random motion' dust particles in a ray of light. It has been observed that the average Fourier power spectrum of the texture image obeys a power law scaling. This fractal property of this motion can be described in terms of the power spectrum of the movement [14].

The image is scanned to obtain a line scan that is formed by an array of light intensities corresponding to this line. Fourier transformation is performed on this array and the power spectrum is computed for each line scan of the image. The power spectra are averaged for all the lines of the image and the log scale plot corresponding to the PSD and the frequency is made. FD is computed as the slope of this plot. However, this suffers from the limitation that it is slow, and has a large number of computational steps.

2.9 Relationship of Fractals and Self-similarity

Nonlinearity is the cause of fractals and chaos and this is explained by the phenomena referred to as stretch and fold. Consider the use of dough to make a pastry which requires repetition of a simple process of stretching the dough followed by folding it. Two points that may have been next to each other at the start would end up distant from each other after multiple layers are formed. This simple technique explains a number of complex phenomena including growth of the body, or the formation of cells [12–14].

2.9.1 Sierpinski triangle

A simple equilateral triangle, when repeated multiple times, leads to an attractive shape, and is called the Sierpinski triangle, named after the discoverer of this pattern. It is also known as Sierpinski Sieve, and has fractal geometry. Starting with an equilateral triangle, it is subdivided recursively into smaller equilateral triangles, which is performed by joining the mid points of each the sides, and if this is continued, the resultant is the sierpinski triangle as shown in Fig. 2.1. This is a simple example of a self-similarity based system.

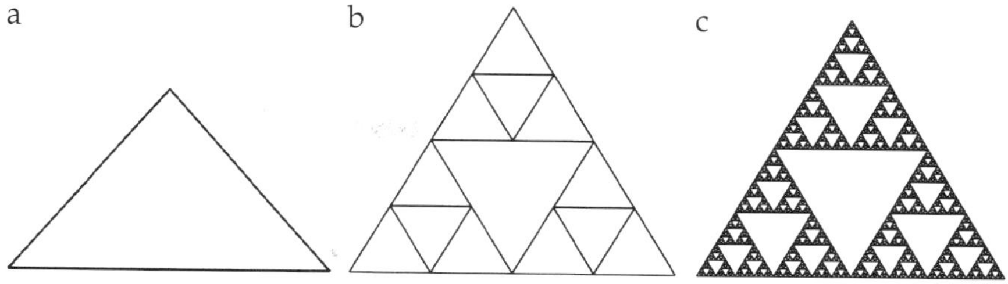

Figure 2.1. Generation of Sierpinski Triangle.

2.9.2 Fractal dimension of the Menger Sponge

A cube is generally considered as a three dimensional body. However, when it is subdivided using an order, such as dividing it as shown in the Fig. 2.2, this limits the freedom of the dimension from 3 to less than 3.

What is the change in the scale or magnification of the cubes scale from left to right? The fundamental, which can be considered as having order 0, has side of L, while the subsequent iteration gives of side L/3 and total of 20 cubes, $N = 20$. Thus scale $r = 3$, and D can be computed as

$$D = \log(N)/\log(r) = \log(20)/\log(3) = 2.726.$$

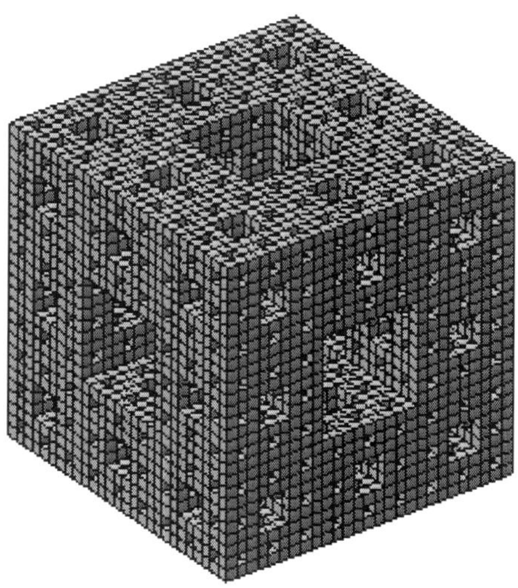

Figure 2.2. Menger Sponge.

2.10 Fractals in Biology

Biological systems, organizations or objects can often be considered to be Fractals. These systems have irregular shape; have spatial or temporal self-similarity, iterative pathways, and a non-integer dimension—the fractal dimension. These also have higher levels of functionality at coarser scales; for example while an individual neuron has simple functions, the brain has the ability to perform very high levels of functions. However, while mathematical fractals are deterministic invariant and self-similar over an unlimited range of scales, biological systems or objects and structures are iterated entities that have statistical self-similarity. This self-similarity is observed in a small window, called scaling window, also called its fractal domain. The fractal dimension of the biological system remains constant within the scaling window. In this section, the relationship between the change in measurement and scale on the log scale remains linear and serves to quantify the variations of the measured quantity with changes in the size of the measuring scale [15].

The biological system is referred to as Fractals when the scaling range has a range of at least two orders of magnitude. The application of the fractal principle is very valuable for measuring dimensional properties and spatial parameters of irregular biological structures, for understanding the architectural/morphological organization of living tissues and organs, and for achieving an objective comparison among complex morphogenetic changes occurring through the development of physiological, pathologic and neoplastic processes.

It is now accepted that the brain has fractal properties. This is true at both; functional and anatomical levels. It is also been found to be an indicator of pathological conditions, and provides a method for describing the complexity of biological systems such as the brain. The complexity and intricacy while being based on very simple basic properties makes this suitable for fractal geometry, though it is beyond standard geometry [4].

2.11 Properties of Natural and Synthetic Objects

Natural objects have fractal properties that have been found to be useful to identify manmade objects in complex natural environment. It has shown that many natural systems have self-similarity and are formed by repetition of similar geometric patterns over multiple scales of observation. The measurement of such patterns can never be done accurately because as the scale is changed, the length changes. There are many examples of this, such as the natural coastline where Euclidean geometry can never measure these exactly and the length can only be estimated at a scale. Repeating the process with a different scale would result in a different measurement.

Objects that are generally man-made are based on calculus and can have the surface or edges defined by polynomial equations. Such manufactured objects would have a smooth edge and lack the roughness, and such objects would have non-fractal geometry. Thus, the non-fractal properties of man-made objects can be considered to identify such objects in natural surroundings. There are number of applications of this property, such as identifying man-made disturbances in forests to identify movement of drugs or weapons by the law enforcement authorities.

This indicates that images of the naturally evolving biological system would display fractal properties while those that are affected by other factors tend to have less fractal properties. This property has been found to be extremely useful for identifying disease conditions, and conditions of ageing, as these do not correspond to an evolving system. Chapter 8 and Chapter 9 describe examples where the fractal properties of retinal vasculature and mammograms have been shown to be an indicator of disease conditions.

2.12 Human Physiology

The chaotic nature of the biological functions has been found to be necessary for the health of the organism. It has been shown that healthy human physiology is best defined in terms of the variability, entropy and fractal dimension rather than in terms of frequencies and time periods [2]. There are number of examples where it has been demonstrated that the physiological parameters can be defined in terms of fractal properties and in most cases, healthy people have higher fractals compared with disease conditions. This may, in general, be attributed to the fractal property described earlier. Systems that are evolving have higher level of chaos and thus the fractal dimension is higher. However, factors such as disease or ageing of the body, which are contrary to evolution, and are result of external causes would have reduced fractal dimension. Over the years, numerous biological parameters and recordings have been analyzed to identify their fractal properties. Below is a brief look at the fractal properties of some of the commonly recorded biomedical signals.

2.12.1 Fractals and Electrocardiogram (ECG), Electromyogram (EMG) and Electroencephalogram (EEG)

Over the past three decades, there has been significant research where fractal dimensions of different biomedical electrical recordings have been obtained. Efforts have been made to consider fractal dimension as a feature and classified against factors such as cardiac activity, muscle force, disease and age [16–18]. Research has found that the ECG of healthy people has higher fractal dimension compared with disease conditions. Results have also identified the fractal properties of EMG associated with the intrinsic properties of the muscles. These have been discussed with examples in later chapters.

2.12.2 Fractal dimension for human movement and gait analysis

While earlier research had suggested that human movement such as our gait is well defined by periodicity, it has now been discovered that human movement is not periodic but chaotic [19]. If any functional was absolutely repetitive, there would be extremely high levels of wear and tear in it, leading to early damage. When human or animals perform repetitive actions, such as walking, these are similar, but small changes are inherent and these prevent injury and fatigue. This can also be seen at the muscle level, where the selection of muscle fibers to produce force is also changed in a chaotic way. These have also been further examined in the later chapters.

2.13 Summary

Chaotic systems are not random although they may appear to be. They are complex but well defined and have some simple defining features:

Chaotic systems are deterministic. This means they have some determining simple function that describes their behaviour.

Chaotic systems are very sensitive to the initial conditions. A very slight change in the starting conditions results in divergence in possible outcomes making these systems fairly unpredictable.

Chaotic systems appear to be disorderly and random, but are ordered. Beneath the seemingly random behaviour is a sense of order and pattern. Truly random systems are not chaotic. The orderly systems predicted by classical physics are the exceptions.

Fractals dimension is a measure of the chaotic properties.

Many biological systems have chaotic properties. In general, healthy biological organisms are more chaotic compared with disease conditions.

References

1. Lorenz, E.N. 1963. Deterministic nonperiodic flow. J. Atmos. Sci., 20: 130–141.
2. Anderson, K.M., P.M. Odell, P.W. Wilson and W.B. Kannel. 1991. Cardiovascular disease risk profiles. Am. Heart J., 121(1 Pt 2): 293–298.
3. Wang, Z. and W.E. Hoy. 2005. Is the Framingham coronary heart disease absolute risk function applicable to Aboriginal people? Med. J. Aust., 182(2): 66–69.
4. D'Agostino, Sr. R.B., S. Grundy, L.M. Sullivan and P. Wilson. 2001. For the CHD Risk Prediction Group. Validation of the Framingham Coronary Heart Disease Prediction Scores: Results of a Multiple Ethnic Groups Investigation. JAMA, 286(2): 180–187.
5. Zmeskal, O., P. Dzik and M. Vesely. 2103. Entropy of fractal systems, Computers & Mathematics with Applications, 66(2): 135–146.
6. Shannon, C.E. 1948. A mathematical theory of communication. Bell Syst. Tech. J., 27: 379–423, 623–656.

7. Rényi, A. 1961. On measures of entropy and information. Proc. Fourth Berkeley Symp. Math. Stat. and Probability, Vol. 1. Berkeley, CA: University of California Press, pp. 547–561.
8. Higashi, M. and G.J. Klir. 1982. Measures of uncertainty and information based on possibility distributions. Int. J. General Syst., 9: 43–58.
9. Lopes, R. and N. Betrouni. 2009. Fractal and multifractal analysis: A review. Medical Image Analysis, 13: 634–649.
10. Russell, D.A., J.D. Hanson and E. Ott. 1980. Dimension of strange attractors. Phys. Rev. Lett., 45: 1175–1980.
11. Sarkar, N. and B.B. Choudhuri. 1994. An efficient differential box counting approach to compute fractal dimension of image. IEEE Transactions on Syst. Man Cybernetics, 24: 115–120.
12. Vaughan, J. and M.J. Ostwald. 2009. Nature and architecture: revisiting the fractal connection in Amasya and Sea Ranch. 43rd Conference of Architectural Science.
13. Cvitanovi, P., R. Artuso, P. Dahlqvist, R. Mainieri, G. Tanner and G. Vattay. 2003. Classical and Quantum Chaos Part I: Deterministic Chaos. Dk/chaosbook.
14. Komulainen, T. 2003. Self-Similarity and Power laws, Helsinki University of Technology (neocybernetics.com/report145/Chapter10.pdf).
15. Losa, G.A. 2013. Fractals and their contribution to biology and medicine, Medicographia, Switzerland.
16. Esgiar, A.N. and P.K. Chakavorty. 2004. ECG signal classification based on fractal features. Conference on Computers in Cardiology, pp. 661–664.
17. Turcott, R. and M.C. Teich. 1996. Fractal character of ECG: Distinguishing heart-failure and normal patients. Annals of Biomedical Engineering, 24: 269–293.
18. Misra, A.K. and S. Raghav. 2010. Local FD based ECG arrhythmia classification. J. Biomedical Signal Processing and Control, 5(2): 114–123.
19. Hausdorff, J. 2004. Gait dynamics, fractals and falls: Finding meaning in the stride to stride fluctuations of human walking. Hum. Mov. Sci., pp. 555–589.

CHAPTER 3

Fractal Dimension of Biosignals

ABSTRACT

Fractal dimension, first formalised in 1975 by Mandelbrot, is an index for quantifying the fractal properties of an event, object or system. Often the terms such as fractal dimension, chaos, complexity and information are used as if interchangeable. What do these terms mean, and what is the relationship between these terms? This chapter explores these terms and the relationship between them in relation to biomedical signals. Over the past three decades, there has been significant research where fractal dimensions of different biomedical electrical recordings and human movement have been computed and associated with disease conditions. Efforts have been made to consider fractal dimension as a feature and classified against factors such as cardiac activity, muscle force, disease and age. Numerous algorithms have been developed that are suitable for measuring the fractal dimensions (FD) of biomedical signals such as ECG, EMG and EEG. While the fundamental concepts are similar, there are significant differences between these methods and the outcomes can also be very different. This chapter discusses some of these algorithms for computing Fractal dimension and the associations with bioelectric recordings and gait analysis.

3.1 Introduction

Fractal geometry and chaos theory provide a new perspective to view physiological signals. While in the world of Calculus mathematics, everything is well defined, and it should be able to accurately predict the future, this is often not possible, and such maths does not respond accurately to real world observations. Many real systems are ordered but unpredictable and are called Chaotic. Fractal geometry is a new language used to describe, model and analyze chaotic and complex forms found in nature. Benoit Mandelbrot was largely responsible in formalising chaos theory and fractals and he showed the presence of fractals in many natural, biological and mathematical descriptions [1].

Fractal, or fractional dimension, is generally used for describing a pattern that repeats itself on an increasingly smaller scale. Alternatively, it can be said that a fractal is a set of self-similar patterns. Mandelbrot [1] observed that certain natural geometries, e.g., coastlines, terrain and clouds, exhibited a simplifying invariance under scale, i.e., their geometries possessed similarities that were invariant to changes in magnification or resolution. He discovered that this invariance to scale existed in a large variety of artificial and natural phenomena. This invariance to scale, i.e., self-similarity, is central to Fractal Geometry, and shows that when a simple process is repeated over and over again at changing scales, the outcome can be astonishingly complex. While most natural phenomena are scale invariant only to a limited scale range, a wide class of natural geometries appears to possess this underlying fractal character within a range of scales.

Let 'F' represent a Fractal. The basic properties of 'F' are [2,3]:

a) F has a fine structure, i.e., detail on arbitrarily small scales.
b) F is too irregular to be described in traditional geometrical language, both locally and globally.
c) F has some self-similarity, perhaps approximate or statistical.
d) Usually Fractal dimension of F is greater than its topological dimension.

Fractals model complex physical processes and dynamical systems. The underlying principle of fractals is that a simple process that goes through infinite iteration becomes a very complex process. Fractal Dimensions are used to measure the complexity of these objects. This chapter will introduce the concepts of fractal dimension, complexity and self-similarity which can be used to analyse the characteristics of bio-signals such as Electrocardiogram (ECG), Electromyogram (EMG), and Electroencephalogram (EEG).

3.2 Fractal Dimension and Self-similarity

The concept of a fractal is most often associated with geometrical objects and the two most famous examples are the Sierpinski triangle and the Koch curve. Both satisfy two important properties:

(i) self-similarity
(ii) fractional dimensions

Mathematically, this property should hold on all scales but in the real world, there are necessarily lower and upper bounds over which such self-similar property applies. The second criterion for a fractal object is that it has a fractional dimension. This requirement distinguishes fractals from Euclidean objects, which have integer dimensions.

As a simple example, a solid cube is self-similar since it can be divided into sub-units or smaller solid cubes that resemble the large cube, and so on. It should be observed that the smaller cubes are similar but not the same as the scale is different. However, while the original cube is not a fractal because it has a dimension = 3 [4,5], the sponge that results is fractal as this limits the number of dimensions of the cube (refer Chapter 2). The same principle applies to the Serpinski triangle (Chapter 2, Fig. 2.1), which, while on a two dimensional plane has dimensionality less than 2.

The concept of a fractal structure, which lacks a characteristic length scale, can be extended to the analysis of complex temporal processes. Although time series are usually plotted on a 2-dimensional surface, it actually involves two different physical variables. The important challenge is in detecting and quantifying self-similar scaling in complex time series [6,7]. An important defining property of a fractal is self-similarity, which refers to an infinite nesting of structure on all scales.

3.2.1 Self-similarity

Self-similarity is a distinctive feature of most fractals. Self-similar processes are the ones in which a small portion of the process resembles a larger section when suitably magnified indicating scale invariance of the process. Self-similarity, in a strict sense, means that the statistical properties of a stochastic process do not change for all aggregation levels of the stochastic process. The stochastic process looks the same irrespective of any magnification of the process. The following will illustrate various types of self-similarity as well as present some real world examples [4,6–8].

Exact self-similarity

Exactly self-similar fractal objects are identical regardless of the scale or magnification at which they are viewed. Strict self-similarity refers to a characteristic of a form exhibited when a substructure resembles a superstructure in the same form. The well-known Koch snowflake curve created by starting with a single line segment and replacing each line segment by four other shapes on iteration as shown in Fig. 3.1 is a good example of the exact self-similar object [4].

Approximate self-similarity

The more common type of self-similarity is the approximate self-similarity. Approximate self-similar objects have recognisably similar object at different scales but are not exactly the same.

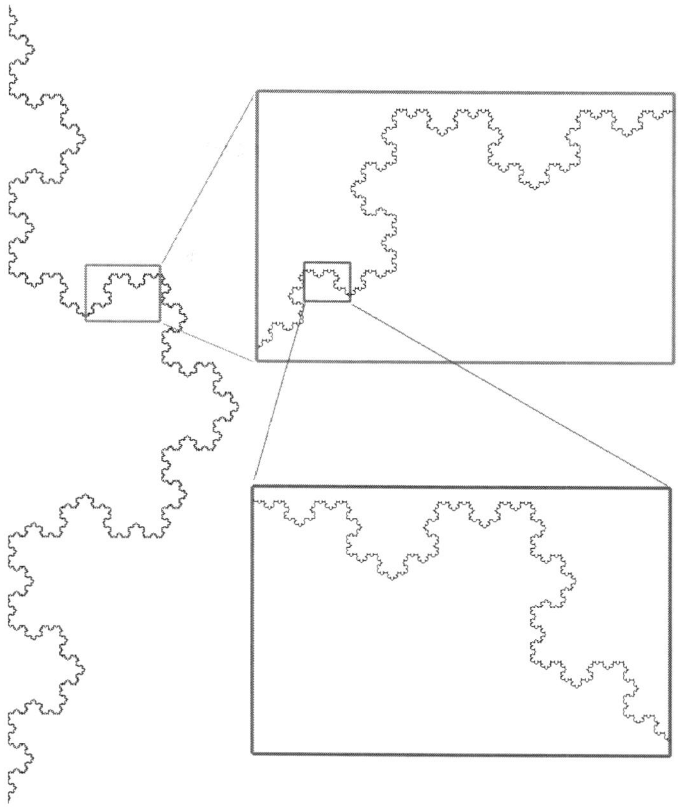

Figure 3.1. Example of exactly self-similar object.

Statistical self-similarity

The self-similar units of a time series signal sometimes cannot be visually observable but there may be numerical or statistical measures that are preserved across scales to determine the self-similar units. This is termed as statistically self-similar. Most physiological signals fall into the category of having statistically self-similar property. An example of statistical self-similar object is 1/f noise (Fig. 3.2), where the units statistically resemble across multiple zooming levels [4].

3.2.2 Fractal dimension

Fractal dimension of a process measures its complexity, spatial extent or its space filling capacity and is related to shape and dimensionality of the process. The concept of fractal can be applied to physiological processes

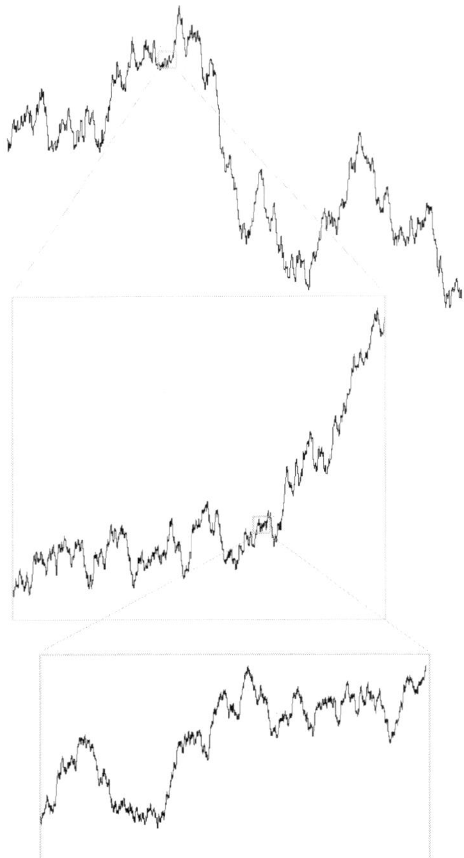

Figure 3.2. Example of statistically self-similar object.

that have self-similar fluctuations over a multiple scale of time and have broad band frequency spectrum [9].

There are many fractal dimensions reported in literature [10–14] including morphological (self-similarity, Hausdorff, mass) and entropy (gyration dimension, information, correlation, variance). The dimension is simply the exponent of the number of self-similar pieces with magnification factor N into which the figure may be broken.

Given a self-similar set S, the fractal dimension D of this set S defined as ln k/ln M where k is the number of disjoint regions that the set can be divided into, and M is the magnification factor of the self-similarity transformation [1,6,7]. This definition of the fractal dimension of a self-similar object is expressed as

$$Fractal\,dimension\quad \frac{\log(number\ of\ self\quad similar\ pieces)}{\log(magnification\ factor)} \tag{3.1}$$

A simple example of computation of fractal dimension of the Sierpinski triangle is illustrated below. Consider the Sierpinski triangle shown in the Fig. 2.1 consisting of 3 self-similar pieces, each with magnification factor 2. So the fractal dimension of this triangle as per the above expression (eqn. 3.1) is

$$Fractal\,dimension = \frac{\log 3}{\log 2} = 1.58 \tag{3.2}$$

Hence the dimension of Sierpinski triangle is between 1 and 2. Fractal dimension is a measure of complexity of a self-similar structure and it measures how many points lie in a given set. A plane is larger than a line, while the dimension of Sierpinski triangle lies in between these two sets [11].

The fractal properties of a time series signal such as Electrocardiogram (ECG), Electroencephalogram (EEG) and Electromyogram (EMG) can also be characterised by computation of fractal dimension. As explained in Section 3.1, the irregularity seen on different scales of time series is not visually distinguishable, it is an observation that can be confirmed by statistical analysis [12,13]. The roughness of the time series signals like biosignals, possesses a self-similar or scale-invariant property and their complexity can be analysed using fractal dimension.

3.3 Different Methods to Estimate Fractal Dimension of a Waveform

In this section, we have discussed the commonly used algorithms and methods for estimation of Fractal dimension of a signal or a waveform which exhibits the characteristics of the fractal structure.

3.3.1 Box-counting method

Box-counting method is one of the most common techniques to measure the fractal dimension of a time-series [1,14]. The basic theory behind this method is covering the object with the small cell of definite size. For example, estimating the box counting Fractal dimension of a 3D object involves counting the number of cubes N(s) of side length s required to cover the object.

$$N(s)\ \alpha\ 1/s^{D} \tag{3.3}$$

where D is the slope of log(N(s)) vs. log(1/s) termed as the *fractal dimension*.

Box counting fractal dimension (BFD) is also known as Kolmogorov dimension or capacity dimension. It is defined as

$$\text{BFD} = \lim_{\epsilon \to 0} \frac{\log \mathcal{M}(\epsilon)}{\log \frac{1}{\epsilon}} \tag{3.4}$$

where $\mathcal{M}(\epsilon)$ denotes number of cells of size ϵ required to cover the waveform such as EMG, EEG and ECG. An estimate of BFD can be obtained by computing the slope of the regression line when log $\mathcal{M}(\epsilon)$ is plotted against log $1/\epsilon$ [15].

3.3.2 Katz's algorithm

According to Katz's algorithm [14–16], the Fractal dimension is obtained directly from the time series and can be generally defined as:

$$\text{Katz Fractal Dimension (KFD)} = \frac{\log L}{\log d} \tag{3.5}$$

where L is the total length of the time-series signal and d is the distance between the first point in the time series and the point that provides the furthest distance with respect to the first point.

For a N-points time series x(1), x(2),...x(N) the KFD can be computed considering L as the sum of distances between successive points, i.e., L = sum (distance (I, I + 1)) for i = 1 to N. Based on this formulation, the KFD is dependent on the particular units of measurement. In order to overcome this problem, Katz introduced a standard variable *a* as a standard unit of measurement which is defined as the average distance between successive points, a = mean (distance (i, i + 1)). Normalisation of the distances was performed by defining

$$n = \frac{L}{a} \tag{3.6}$$

where n is the number of steps in the series.

Using eqn. (3.5) and eqn. (3.6) normalised KFD can be defined as

$$\text{KFD}_{\text{norm}} = \frac{\log n}{\log {}^{d}\!/\!_{L} + \log n} \tag{3.7}$$

3.3.3 Higuchi's algorithm

Higuchi algorithm [17,18] has been used to estimate the Fractal dimension for non-periodic and irregular time series signals. This algorithm yields a more accurate and consistent estimation of FD for physiological signals than other algorithms [18].

The first step for computing the FD requires the computation of the length of the curve, X_k^m, for a time signal sampled at a fixed sampling rate, $x(n) = X(1), X(2), X(3),\ldots\ldots, X(N)$ as follows:

$$L_m(k) = \frac{\left\{\left(\sum_{i=1}^{\left[\frac{N-m}{k}\right]} \left|X(m+ik) - X(m+(i-1).k)\right|\right)\frac{N-1}{\left[\frac{N-m}{k}\right].k}\right\}}{k} \tag{3.8}$$

where [ ] denotes the Gauss' notation and both k and m are integers. m = initial time; k = time interval; i = 1 to $N\left[\frac{N-m}{k}\right]$

The term $\frac{N-1}{\left[\frac{N-m}{k}\right].k}$ represents the normalization factor for the curve length of subset time series. The length of the curve for the time interval k,$\langle L(k)\rangle$ is defined as the average value over k sets of $L_m(k)$. If $\langle L(k) \propto k^{-D}\rangle$, then the curve is fractal with the dimension D.

3.3.4 Petrosian's algorithm

Petrosian's Algorithm provides a fast computation and quick estimate of the FD [14,8,19]. This estimate is computed by translating the time series into a binary sequence. Considering all waveforms are analog signals, a binary signal is derived as the following five different methods denoted with the letters *a, b, c, d* and *e* [18,19].

- Method *a* generates the binary sequence by assigning ones when the waveform value is greater than the mean of the data window under consideration, and zero when it is lower than the mean.
- Method *b*, the binary sequence is formed by assigning one each time the waveform value is outside the band of the mean plus and minus the standard deviation, and assigning zero otherwise.
- Method *c* constructs the binary sequence by subtracting consecutive samples on the waveform. From this sequence of subtractions, the

binary sequence is created by assigning +1 or –1 depending on whether the result of the subtraction is positive or negative respectively.

- In method *d*, the differences between consecutive waveform values are given the value of one or zero depending on whether their difference exceeds or not a standard deviation magnitude.
- Method *e* is a variation of method d which consists of utilizing an a priori chosen threshold magnitude different from the standard deviation.

The Petrosian's fractal dimension (PFD) by any of the methods mentioned above is then computed as

$$\mathrm{PFD} = \frac{\log n}{\log n + \log\left(\dfrac{n}{n + 0.4N_{\Delta}}\right)} \tag{3.9}$$

where n is the length of the sequence (number of sample points) and N_{Δ} is the number of sign changes in the binary sequence.

3.3.5 Sevcik's algorithm

Sevcik's algorithm [14,20] estimates fractal dimension from a set of N values, the series y_i sampled from the waveform between time 0 and t_{max}. Then double linear transformation is performed on the waveform that maps it into a unit square. The normalised abscissa of the square is x_i^* and the normalised ordinate is y_i^* and is defined as follows:

$$x_i^* = \frac{x_i}{x_{max}} \tag{3.10}$$

$$y_i^* = \frac{y_i - y_{min}}{y_{max} - y_{min}} \tag{3.11}$$

where

x_{max} = maximum of x_i

y_{max} and y_{min} = maximum and minimum of y_i

Sevcik fractal dimension (SFD) of the waveform is then approximated as

$$SFD = 1 + \frac{\ln(L) + \ln(2)}{\ln(2N')} \tag{3.12}$$

where L is the length of the time series in the unit square and N' = N – 1.

3.3.6 Correlation dimension

Correlation dimension is used as estimate to measure dimension of fractal objects [14]. It is computed using Grassberger-Procaccia algorithm [21] and based on the Theiler method [22]. This algorithm considers spatial correlation between pairs of points on a reconstructed attractor. Consider N number of points in a waveform denoted by $\mathbf{x}_1 \ldots\ldots\ldots\ldots \mathbf{x}_N$ in some metric space with distances $|\mathbf{x}_i - \mathbf{x}_j|$ between any pair of points. For any positive number r, the correlation sum $C(r)$ is then defined as the fraction of pairs, whose distance is smaller than r,

$$\hat{C}(r) = \frac{2}{N(N-1)} \sum\nolimits_{i<j} \theta\left(r - \left|\mathbf{x}_i - \mathbf{x}_j\right|\right) \tag{3.13}$$

where $\theta(x)$ is the Heaviside step function. It is an unbiased estimator of correlation integral

$$C(r) = \int d\mu(\mathbf{x}) \int d\mu(\mathbf{y}) \theta(r - |\mathbf{x} - \mathbf{y}|) \tag{3.14}$$

Both $C(r)$ and $\hat{C}(r)$ are monotonically decreasing to zero as $r \rightarrow 0$. If $C(r)$ decreases like a power law, $C(r) \sim r^D$, then D is termed as the correlation dimension [14].

3.3.7 Adapted box fractal dimension

The Adapted Box fractal dimension (ABFD) was introduced by Henderson et al. [23–25]. The dimension of a waveform can be estimated by dividing a time series of T durations into segments of length Δt. The difference between the minimum and maximum duration is derived for each segment, which is termed as the extent. The mean extent $E(\Delta t)$ is then computed for a range of Δt and the dimension is computed by finding the best fit to the following equation [14]:

$$A(\Delta t) = TE(\Delta t) \approx A_0 \, \Delta t^{2-D} \tag{3.15}$$

where T is the total duration of the series, E is the mean extent, Δt is the segment length and D is the estimated ABFD.

3.3.8 Fractal dimension estimate based on power law function

The power spectrum (the square of the amplitude from the Fourier transform) of a pure fractional Brownian motion is known to be described by a power law function [26,27]:

$$|A|^2 \approx 1f^{\beta} \tag{3.16}$$

where $|A|$ is the magnitude of the spectral density at frequency f, with an exponent equal to

$$\beta = 2H + 1 \tag{3.17}$$

In general, fractal signals always have a very broad spectrum. When the derivative is taken from a fractal signal, β is reduced by two. Hence for a fractional Brownian noise, fBn, β is expected to be: β = 2H – 1. β is the slope derived by fitting a straight line in the logarithmic plot and H is calculated from the slope β. The fractal dimension relates to H as D = E + 1 – H, where E is the Euclidean dimension [27].

3.4 Fractals and Electrocardiogram (ECG), Electromyogram (EMG) and Electroencephalogram (EEG)

The analysis of time-series signals such as ECG, EMG and EEG has been an important part in today's world of health and technology. Most bio-signals show repetitive patterns which provide unique information in terms of the clinical diagnosis. It is important to analyse these specific patterns and its relation with the underlying physiological processes.

Traditional techniques such as time series analysis and frequency analysis have been applied to provide better insight to show the changes in the periodicity of the signal. But these analyses showed that the variations were not exact since the exact periodicity of cardiovascular events denotes an abnormal phenomenon [28]. This is because biological control systems are multi-dimensional, low-gain, and has lot of redundant information.

In past decade, research studies have shown the presence of non-linearity in biological signals as also, considering the signals as chaotic dynamical systems [29]. In order to investigate the chaotic dynamic system, fractal theory brings in more useful information to study the more natural and repeated patterns in the signal. Based on these studies, examining the fractal or chaotic nature of a signal is valuable, in order to understand how the system works and develop new techniques for analysis.

Fractal systems can be analysed in time or space domain. These systems have patterns which correlate at all levels of scale. Fractal analysis has been used for analysing the characteristics of cardiac (ECG) signal. Research studies have shown that ECG signals can be modelled as fractal sets and can be classified using fractal dimension [28–32]. Peng et al. [33] studied scale-invariant properties of the human heartbeat time series which is termed as the output of a complicated integrative control system.

The sequences of heartbeats show fluctuations which have scale-invariant property [30–32]. These fluctuations can be analysed to observe the presence of repeated patterns and can be associated with various pathological conditions, such as diabetic autonomic neuropathy, heart disease and myocardial infarction. The changes in the repeated patterns in various level of scale can be related to a particular condition based on the normal heart rhythms. Example of the fractal analysis of ECG has been explained in Chapter 4.

Fractal theory has been widely used for analysis of Electromyogram (EMG) signal and its applications. Electromyogram signal is the recording of the muscle activity and is the combination of the muscle fiber action potentials which can be detected by a sensor located on the surface of the skin. Fractal dimension of sEMG has been found sensitive to magnitude and rate of muscle force generated. The fractal dimension is introduced as the index for describing the irregularity of a time series in place of the power law index. Gitter et al. [34] demonstrated that the fractal characteristics of EMG signal with a dimension is highly correlated with muscle force.

EMG signal is a result of the summation of identical motor units that travel through tissues and undergo spectral and magnitude compression. Burst behaviour of sEMG in time has the property that patterns observed at one sampling rate are statistically similar to patterns observed at lower sampling rates. These patterns suggest that sEMG has self-similarity property [35]. The changes in the properties of the sEMG signal have been observed due to age, fatigue and other neuromuscular disorders. Aging is considered to have a strong impact on the parameters of muscle functions such as strength, endurance, and fatigue. There is associated decrease in muscle mass, caused by loss of muscle fibres numbers and decrease in muscle fibre sizes [36], leading to diminished muscle function. This decrease may be a cause of reduction in the complexity in the sEMG signal and can be measured using fractal dimension. An example of the fractal analysis of sEMG has been explained in Chapter 5.

Electroencephalogram (EEG) is the recording of the electrical activity of a brain which exhibits a significant complex behaviour with strong non-linear and dynamic properties. EEG signals are highly non-Gaussian, non-stationary, complex and random in nature. The characteristics of EEG mainly depend on the age, mental state and vary between individuals. The processing of EEG signals involves linear and non-linear techniques as it is important to understand its correlation to the physiological events. The occurrence of the events in the brain is random in nature and requires non-linear techniques for correlation. In order to study these underlying complex characteristics of brain which is reflected in EEG, fractal theory has been investigated by many researchers.

One of the earlier studies [37], reported that the EEG exhibits fractal like behaviour which suggests that the human brain is fractal in time. The fractal dimension of EEG signal which measures the random and complex nature can be correlated with the different states of the brain activity. Example of the fractal analysis of sEMG has been explained in Chapter 6.

3.5 Fractal Dimension for Gait Analysis

Gait is the pattern of the movement during human walking. The analysis of gait has been widely used to obtain information about various musculoskeletal and neurological conditions. During gait, the locomotion is based on the one stride after the next in a regular fashion. There are two important phases in the walking pattern: *Stance Phase* and *Swing Phase*

- *Stance phase* is the time when the foot is on the ground and occupies about 60% of the walking cycle. At some period of time in this phase, both feet will be on the ground for a period of time.
- *Swing phase*: In this phase, one foot is on the ground and one in the air.

The gait cycle is termed to be a repetitive pattern with steps and strides and can show self-similar pattern. The transition in the phases, fluctuations and the interactions that occur in the gait cycle happen at the microscopic level. This gait rhythm is known to have a fractal property and the presence of the fractal rhythm in gait was reported based on following reasons [38,39]:

a) the presence of a non-trivial long-term dependence effect and over the long-term the fluctuations, i.e., stride–stride variations are random
b) fluctuations are often associated with a non-equilibrium dynamical system with multiple-degrees-of freedom
c) on an average, the fluctuations in the stride interval are apparently related to variations in the stride interval hundreds of strides earlier, which exhibits scale-invariance.

Research studies have reported that the stride interval time series have been identified as to exhibit self-similarity across all scales of magnification and considered as the fractal sets [39]. The stride interval time series can be characterized by fractal pattern depend upon several biological and stress constraints. When healthy young adults walk slowly, the fractal rhythm is not reduced. The fractal gait rhythm or pattern is changed when one is asked to walk in different timing instead of their own natural timing mechanism to walk. There is a linear change in fractal dimension even in healthy subjects who walk with and without constraints.

The fractal dynamics of the stride interval is observed to be altered with advanced age and disease. The changes in the fractal rhythm in these populations are not only attributable to reduced gait speed or increased stride-to-stride variability but also depend on some aspect of the neuromuscular control system that is not directly related to walking velocity or gait unsteadiness [40]. Abnormalities in human walking are usually caused by diseases or injuries in the legs, feet or brain. Fractal dimension analysis can be used to identify patients with gait disorders by observing their walking rhythm.

Balance control is one of the stances in the human locomotion where the control varies due to age or disease. Center of pressure (CoP) trajectory where the length, surface, maximal amplitude of the displacement, speed, and frequency is analysed. CoP provides a measure of whole body dynamics and thereby represents the sum of various neuro-musculoskeletal components acting at different joints level. The variations in the CoP can be quantified using Fractal dimension as it may indicate a change in control strategies for maintaining quiet stance [41].

Research studies [42–44] have reported that the fractal analysis represents a reliable method to highlight specific characteristics of balance control. Doyle et al. [43] assessed the reliability of traditional and fractal dimension measures of quiet stance CoP in young healthy individuals. Blaszczyk et al. [44] used fractal dimension technique in healthy elderly individuals with their eyes open and closed and reported that there is a change in fractal dimension resulting from the change in stability and balance.

3.5.1 Example

The disturbances in gait patterns show symptoms of disability and impairment in patients with neuromuscular disorders and Parkinson's disease (PD). Gait hyperkinesia is the first to be observed in PD patients and it is responsible for the impairment in stride length, gait variability, walking velocity and fractal scaling. In addition to these time-distance parameters kinematic and kinetic gait variables, abnormal fractal dimension and entropy play an important role in characterizing gait in PD patients. Research study conducted in Parkinson and ataxia patients [45] have shown that the fractal dimension was more sensitive than traditional stabilometric analysis in the evaluation of postural instability.

Freezing of Gait (FoG) is a symptom of PD in advanced stages where the feet are glued to the floor and there is no initiation of gait [46]. The underlying mechanism of FoG is still unclear and its diagnosis is complex [47,48]. Due to the daily, unpredictable and frequent occurrence of FoG events, they can drastically reduce the quality of life in patients with advanced PD. Neuromuscular and social consequences of FoG events significantly degrade the quality of life in patients with advanced PD.

We investigated the publicly available data [49] collected from the accelerometer sensor located in the shank from the PD patients where they experienced the Freezing of Gait (FoG). The fractal dimension of the signal was computed using Higuchi's algorithm as mentioned in Chapter 3, Section 3.3 during the normal walking and when they experienced FoG. Figure 3.3 shows the plot of FD of accelerometer signal during normal walk and FoG for a PD patient. The results show that the there is a decrease in Fractal dimension during FoG when compared with normal walking of the PD patients. The change denotes that this change is due to the alterations in the gait initiation. The fractal gait rhythm can be used to identify the symptoms of various gait related disorders and aging.

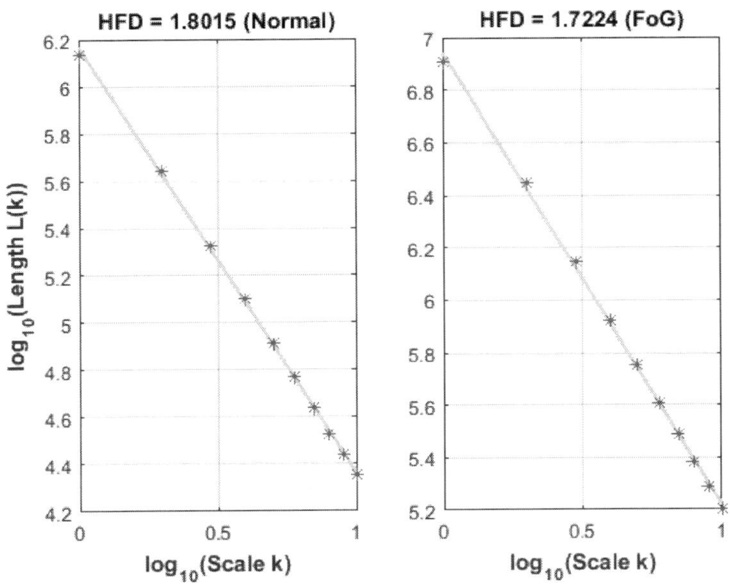

Figure 3.3. Box counting Fractal dimension of accelerometer signal during normal walking and FoG for a PD patient.

3.6 Summary

The chapter has discussed the basics of fractal theory and its properties related to bio-electrical signals and signals corresponding to human movement. It has been demonstrated that fractal dimension is one of the important measures to observe the complex real-world signals such as the bioelectric recordings; electrocardiogram (ECG), Electroencephalogram (EEG) and electromyogram (EMG) and recordings of the human gait. There are number of different algorithms that have been developed to estimate the Fractal dimension of such signals. And these have been presented in this chapter.

Analysis has shown that these signals are fractal and research has shown that changes to FD of these signals can be associated with the changes in the physiological and health parameters. The fractal analysis has also shown that recordings of the human movement are also fractal and this is associated with disease and aging.

References

1. Mandelbrot, B.B. 1977. Fractals: Form, chance, and dimension, first edn., W.H. Freeman and Co., San Francisco.
2. Akujuobi, C. and A. Baraniecki. 1992. Wavelets and fractals: a comparative study. IEEE Sixth SP Workshop on Statistical Signal and Array Processing, 1992. pp. 42–45.
3. Falconer, K. 1990. Fractal Geometry—Mathematical Foundations and Applications, John Wiley and Sons, New York.
4. Bourke, P. 2007. Self similarity, Fractals, Chaos. URL: http://paulbourke.net/fractals/fracdim/ Last Access: 18 March 2016.
5. Feder, J. 1988. Fractals, New York: Plenum Press.
6. Goldberger, A.L., L.A.N. Amaral, L. Glass, J.M. Hausdorff, P.C. Ivanov, R.G. Mark, J.E. Mietus, G.B. Moody, C.-K. Peng and H.E. Stanley. 2000. PhysioBank, PhysioToolkit, and PhysioNet: Components of a new research resource for complex physiologic signals. Circulation 101(23): e215–e220. Circulation Electronic Pages: http://circ.ahajournals.org/cgi/content/full/101/23/e215. Last Access: 18 March 2016.
7. Bassingthwaighte, J., L. Liebovitch and B. West. 1994. Fractal Physiology, Oxford University Press, New York.
8. Iannaconne, P. and M. Khokha. 1996. Fractal Geometry in Biological Systems: An Analytical Approach, CRC Press, Boca Raton.
9. Gupta, V., S. Suryanarayanan and N.P. Reddy. 1997. Fractal analysis of surface EMG signals from the biceps. International Journal of Medical Informatics, pp. 185–192.
10. L´evy-V´ehel, J. and E. Lutton. 2006. Fractals in Engineering: New Trends in Theory and Applications, Springer-Verlag New York, Inc., Secaucus, NJ, USA.
11. Devaney, R.L. 1995. Chaos in the classroom, Mathematics and Statistics at Boston University.
12. Kobayashi, M. and T. Musha. 1982. 1/f fluctuation of heartbeat period. Biomedical Engineering, IEEE Transactions on BME-29(6): 456–457.
13. Peng, C.-K., J. Hausdorff and A. Goldberger. 1999. Fractal mechanisms in neural control: Human heartbeat and gait dynamics in health and disease. Nonlinear Dynamics, Self-Organization, and Biomedicine.
14. Goh, C., B. Hamadicharef, G. Henderson and E. Ifeachor. 2005. Comparison of Fractal Dimension Algorithms for the Computation of EEG Biomarkers for Dementia. 2nd International Conference on Computational Intelligence in Medicine and Healthcare (CIMED2005), Jun. 2005, Lisbon, Portugal.
15. Cusenza, M. 2011. Fractal Analysis of the EEG and Clinical Applications, PhD Thesis, Retrieved from https://www.openstarts.units.it/dspace/bitstream/10077/7394/1/cusenza_phd.pdf. Last Access: 18 March 2016.
16. Katz, M. 1988. Fractals and the analysis of waveforms. Comput. Biol. Med. 1988, 18(3): 145–156.
17. Higuchi, T. 1988. Approach to an irregular time series on the basis of the fractal theory. Phys. D, 31(2): 277–283.
18. Esteller, R., G. Vachtsevanos, J. Echauz and B. Litt. 2001. A comparison of waveform fractal dimension algorithms. Circuits and Systems I: Fundamental Theory and Applications, IEEE Transactions on Circuits and Systems I, 48(2): 177–183.

19. Petrosian, A. 1995. Kolmogorov complexity of finite sequences and recognition of different preictal EEG patterns. Proceedings of the Eighth IEEE Symposium on Computer-Based Medical Systems, Lubbock, TX, pp. 212–217.
20. Sevcik, C. 1998. A procedure to estimate the fractal dimension of waveforms. Complexity Int., 5: 1.
21. Grassberger, P. and I. Procaccia. 1983. Measuring the strangeness of strange attractors. Physica D, 9: 189–208.
22. Theiler, J. 1987. Efficient algorithm for estimating the correlation dimension from a set of discrete point. Physical Review A, 36(9): 4456–4462.
23. Henderson, G.T., E.C. Ifeachor, H.S.K. Wimalartna, E.M. Allen and N.R. Hudson. 2000. First IEE International Conference on Advances in Medical Signal and Information Processing (MEDSIP'00), 284–289.
24. Henderson, G.T., E.C. Ifeachor, H.S.K. Wimalartna, E.M. Allen and N.R. Hudson. 2002. Proceedings of the 4th International Workshop on Biosignal Interpretation, Como, Italy, 319–322.
25. Henderson, G.T., P. Wu, E.C. Ifeachor and H.S.K. Wimalaratna. 1998. Proceedings of the 3rd International Conference on Neural Networks and Expert Systems in Medicine and Healthcare (NNESMED'98), 322–330.
26. Voss, R.F. 1988. pp. 21–70. *In*: Peitgen, H.O. and D. Saupe (eds.). Fractals in Nature: From Characterization to Simulation. Springer-Verlag; New York: 1988.
27. Schepers, H.E., J.H.G.M. van Beek and J.B. Bassingthwaighte. 2002. Four Methods to Estimate the Fractal Dimension from Self-Affine Signals. IEEE engineering in medicine and biology magazine: the quarterly magazine of the Engineering in Medicine & Biology Society. 2002, 11(2): 57–64. doi:10.1109/51.139038.
28. Turcott, R.G. and M.C. Teich. 1996. Fractal character of the electrocardiogram: Distinguishing heart-failure and normal patients. Annals of Biomedical Engineering, 24(2): 269–293.
29. Turcott, R.G., P.D.R. Barker and M.C. Teich. 1995. Long-duration correlation in the sequence of action potentials in an insect visual interneuron. J. Statist. Comput. Simul., 52: 253–271.
30. Goldberger, A.L., D.R. Rigney and B.J. West. 1990. Chaos and fractals in human physiology. Sci. Am., 262(2): 42–49.
31. Freeman, R., J.P. Saul, M.S. Roberts, R.D. Berger, C. Broadbridge and R.J. Cohen. 1991. Spectral analysis of heart rate in diabetic autonomic neuropathy. Arch. Neurol., 48: 185–190.
32. Bigger, J.T., Jr., J.L. Fleiss, R.C. Steinman, L.M. Rolnitzky, R.E. Kleiger and J.N. Rottman. 1992. Frequency domain measures of heart period variability and mortality after myocardial infarction. Circulation, 85: 164–171.
33. Peng, C.K., S. Havlin, H.E. Stanley and A.L. Goldberger. 1995. Quantification of scaling exponents and crossover phenomena in non-stationary heartbeat time series. Chaos, 5: 82–87.
34. Gitter, J.A. and M.J. Czerniecki. 1995. Fractal analysis of the electromyographic interference pattern. J. of Neuroscience Methods, pp. 103–108.
35. Anmuth, C.J., G. Goldberg and N.H. Mayer. 1994. Fractal dimension of electromyographic signals recorded with surface electrodes during isometric contractions is linearly correlated with muscle activation. Muscle & Nerve, 17(8): 953–954.
36. Kent-Braun, J.A. and A.V. Ng. 1999. Specific strength and voluntary muscle activation in young and elderly women and men. J. Appl. Physiol. 1999, 87: 22–29.
37. Pritchard, W.S. 1992. The brain in fractal time: 1/f-like power spectrum scaling of the human electroencephalogram. Int. J. Neurosci. 1992, 66: 119–29.
38. Hausdorff, J.M. 2007. Gait dynamics, fractals and falls: Finding meaning in the stride-to-stride fluctuations of human walking. Human Movement Science, 26(4): 555–589.
39. Chau, T. 2001. A review of analytical techniques for gait data. Part 1: fuzzy, statistical and fractal methods. Gait & Posture, 13(1): 49–66.

40. Rhea, C.K., A.W. Kiefer, S.E. D'Andrea and S.H. Warren. 2014. Aaron, Entrainment to a real time fractal visual stimulus modulates fractal gait dynamics. Human Movement Science, August 2014, 36: 20–34.
41. Hausdorff, J.M., Y. Ashkenazy, C.-K. Peng, P.C. Ivanov, H.E. Stanley and A.L. Goldberger. 2001. When human walking becomes random walking: fractal analysis and modeling of gait rhythm fluctuations. Physica A: Statistical Mechanics and its Applications, Volume 302, Issues 1–4, 15, pp. 138–147.
42. Cimolin, V., M. Galli, C. Rigoldi, G. Grugni, L. Vismara, L. Mainardi and P. Capodaglio. 2011. Fractal dimension approach in postural control of subjects with Prader-Willi Syndrome. J. of NeuroEngineering and Rehabilitation, 8: 45.
43. Doyle, T.L., R.U. Newton and A.F. Burnett. 2005. Reliability of traditional and fractal dimension measures of quiet stance center of pressure in young, healthy people. Arch. Phys. Med. Rehabil. 2005, 86: 2034–2040.
44. Blaszczky, J.W. and W. Klonowsky. 2001. Postural stability and fractal dynamics. Acta Neuro. Exp. 2001, 61: 105–12.
45. Manabe, Y., E. Honda, Y. Shiro, K. Sakai, I. Kohira, K. Kashihara, T. Shohmori and K. Abe. 2001. Fractal dimension analysis of static stabilometry in Parkinson's disease and spinocerebellar ataxia. Neurol. Res. 2001, 23: 397–404.
46. Jankovic, J. 2008. Parkinson's disease: clinical features and diagnosis. J. Neurol. Neurosurg. Psychiatry, 79(4): 368–376.
47. Nutt, J.G., B.R. Bloem, N. Giladi, M. Hallett, F.B. Horak and A. Nieuwboer. 2011. Freezing of gait: moving forward on a mysterious clinical phenomenon. Lancet Neurol., 10(8): 734–744.
48. Plotnik, N., Giladi, Y. Balash, C. Peretz and J.M. Hausdorff. 2005. Is freezing of gait in Parkinson's disease related to asymmetric motor function?, Ann. Neurol., 57(5): 656–663.
49. Daphnet Freezing of Gait Data Set. [Online]. Available: https://archive.ics.uci.edu/ml/datasets/Daphnet+Freezing+of+Gait. Last Access: 18 March 2016.

CHAPTER 4

Fractals Analysis of Electrocardiogram

ABSTRACT

Often the heart activity is approximated to be periodic, and average rate is computed by taking the average of the time duration over a few samples. However, it is now well known that heart activity is not periodic and healthy heart requires continually changing time gap between successive heart beats. This chapter discusses the relationship of fractal properties of heart activity from the view of health and wellbeing. The ancient medicine from the east knew that it was not heart rate but the heart rate variability that was a good indicator of health. The Ayurvedic doctor or Chinese medicine practitioners study the steadiness of the change of the 'pulse' to identify disease. Modern medicine has discovered this phenomenon and HRV or heart rate variability is now routinely used to identify disease. Poincaré, named after Henri Poincaré, is a measure of self-similarity in an array, process or signal and Poincaré plots are being used for quantifying ECG recordings, specially the recordings taken over extended periods of time. Non-linear analysis of ECG that are based on its fractal properties have been found useful in quantifying the recordings and are becoming very useful, especially when considering recordings over extended periods of time.

4.1 Introduction

Healthy functioning of the cardiovascular system is necessary for a healthy body. The cardiovascular system consists of the heart at the centre with a network of vessels and serves the purpose of transporting blood to and from all parts of the body. The heart has no backups or alternates therefore it is necessary for the heart to function properly over the entire duration of the life of the body.

Cardiovascular disease is the single largest cause of human mortality [1]. Understanding the heart activity is important for diagnosis of various cardiovascular and heart related diseases. The study of the heart has now become the single most important scientific discipline and numbers of modalities have been developed to diagnose heart ailments. These include imaging modalities such as ultrasound imaging and Doppler that identify the anatomical details and investigate the flow of blood. However, the most convenient and informative method of investigating the functioning of the heart is based on the electrical recording associated with the heartbeat, and this is commonly referred to as the electrocardiogram (ECG).

The heart performs cyclic activity, and each cycle is referred to as the heartbeat, with the cycle being approximately 1 heartbeat per second. Each beat results in a cyclic electrical signal being generated and this can be recorded using needles or from the surface, and is based on the physiology of the heart. The heart consists of four chambers as explained below [2]:

- The upper two chambers—left and right atria where the blood enters into the heart,
- The lower two chambers—left and right ventricles are contraction chambers sending blood through the circulatory system.

The chambers are of significantly different size and there is a big difference in the duration of the contraction of each of the chambers. There is also the asymmetry of the heart with relation to the body exterior, and thus the recorded electrical activity has a very unique shape.

The cardiovascular circulation is split into two major loops; one passing through the lungs which are termed as pulmonary and other through the body termed as systemic [2]. This provides the ability for the body to function, and for cellular metabolism to take place, supported by the flow of blood. One important factor of the heart activity is that it has to be consistent and regular (not periodic) over the life of the body, and there is no backup for providing rest and recovery opportunity. Thus, it is essential for all the components to be functional despite decay and possible changes within the system.

The heart activity over the cardiac cycle refers to a complete heartbeat. Each heartbeat has several stages of filling and emptying of the four chambers and associated vessels. The frequency of the cardiac cycle is reflected as heart rate and termed as beats per minute (bpm). The heart muscle is self-exciting and operates automatically when compared with other muscles in the body as these require nervous stimuli for excitation. The contractions in the heart occur spontaneously and are rhythmic but are sensitive to large number of intrinsic and extrinsic factors such as temperature and movement. The cardiac activity is influenced by hormones for both; sympathetic and parasympathetic activity [3]. Thus, there are large numbers of parameters that influence the cardiac activity and the activity can best be described as chaotic and the rhythmic contractions have been found to be fractal [4]. Modeling has shown that while periodic and regular cardiac activity would lead to fatigue of the vessels and the muscles, and causing weak points in the length of the vessels, the chaotic nature of the activity would prevent the buildup of fatigue which would otherwise happen because of the regularity. This appears to be the body's inherent ability to prevent fatigue of the cardiac system.

4.1.1 Recording cardiac activity

The heart can be considered as one large muscle, and the contraction of the heart is associated with electrical activity and mechanical movement. The cardiac activity can be monitored using either of the two. The electrical potential can be recorded from the surface of the body using simple electrodes and an amplifier, and is called the electrocardiogram (ECG). The mechanical activity can be monitored using either sensors to monitor the blood flow, or movement of the chest, or the changes to the electrical properties of the chest. The two most common methods used to monitor the cardiac activity are; electrocardiogram (ECG) and Photo-plethysmogram and these are described below [3]:

a) **Electrocardiogram (ECG, EKG).** The electrical potential corresponding to the chain of activity of heart muscle depolarisations are recorded as ECG from the surface of the body. The amplitude is very small but this can be reliably recorded with surface electrodes attached to the skin and amplified using amplifiers with high gain (>100), high common mode rejection ratio (CMRR) and band-pass filters (0.5 to 150 hz). There are a number of different setups suitable for recording ECG, though the most common is where there are three electrodes located such that the ground is near the leg, and the differential are across the thorax. The setup comprises of electrodes located on the chest around the heart or at the four extremities according to standard nomenclature (RA = right arm; LA = left arm; RL = right leg; LL = left leg).

b) **Photo-Plethysmography (PPG).** Each heartbeat causes the blood to be pumped through the blood vessels, where there is a head created in the arteries and there is the return blood to the heart via the veins. The largest chamber of the heart, the left ventricle, pumps the blood through the aorta to the body, generating a pressure wave that can be observed superficially in number of locations in the body. Based on the physiological action leading to pumping in the cardiac cycle, the blood pressure throughout the body changes cyclically; increases and decreases and this cyclic change can be easily superficially monitored. Peripheral blood flow can be measured using optical sensors attached to the fingertip, the ear lobe or other capillary tissue. The device has an LED that transmits light on the skin and the receiver records either the transmitted or reflected light and thus measures how much light is either absorbed or reflected to the photodiode. While not as accurate as ECG recordings, PPG clips are much less intrusive as they are suitable for dry sensors, require a single point of contact and can be attached with less preparation time than compared to ECG recording.

Electrocardiogram (ECG): The heart activity is routinely monitored using ECG. The signal can be considered to be pseudo periodic and composed of five distinct wave shapes; P, Q, R, S, and T waves. Change in the shapes of the signal and waves can reveal disease conditions [5]. For example, large Q wave and inverted T wave refer to pulmonary embolism. A sample of ECG signal is shown in Fig. 4.1.

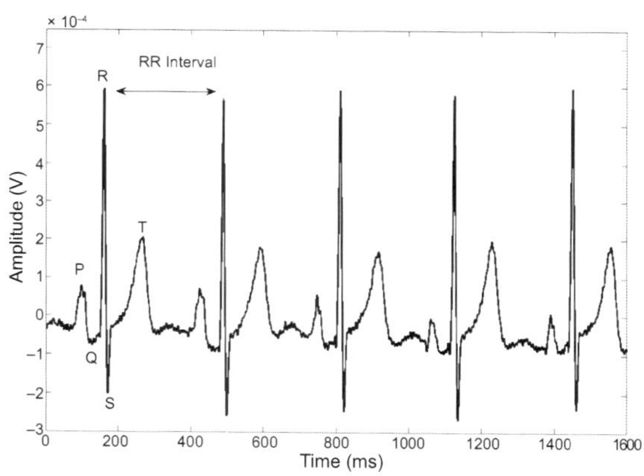

Figure 4.1. Sample ECG Signal and computation of RR interval.

The rhythmic activity in the heartbeat, while appearing very regular and periodic, is however chaotic in nature. It has also been demonstrated to have scale-invariant fluctuations and exhibit fractal behaviour. This fractal behaviour has been reported in various research studies to classify between normal and pathological conditions. Goldeberg et al. [6] has reported that the presence of long-range (fractal) correlations in cardiovascular fluctuations in health has implications for understanding and modeling the neuroautonomic regulation.

Most of these fractal characteristics of heart rate have been assumed as monofractals (e.g., Koch Curve, Fig. 3.2). Monofractals are where it is homogeneous, exhibiting the same scaling properties characterized by one singularity exponent throughout the entire signal [7–8]. Recent research has shown that the heart rate variability can be modelled as *multi* fractal, requiring a larger number of indices to characterize their scaling properties [6,9].

4.2 Heart Rate Variability

Heart rate variability (HRV) is termed as the variation over time of the period between consecutive heartbeats. HRV is predominantly dependent on the extrinsic regulation of the heart rate (HR). Hear rate variability is one of the important parameters in diagnosing the heart related diseases and it reflects the heart's ability to adapt to changing circumstances by detecting and quickly responding to unpredictable stimuli based on autonomic nervous system. HRV analysis is the ability to assess overall cardiac health and useful signals for understanding the status of the ANS [10].

The variability in HR is attributable to the autonomic neural regulation of the heart and the circulatory system [10]. The balancing action of the sympathetic nervous system (SNS) and parasympathetic nervous system (PNS) branches of the ANS control the HR. Increased SNS or diminished PNS activity results in cardio-acceleration. However, a low SNS activity or a high PNS activity causes cardio-deceleration [10].

HRV can be affected due to various factors such as:

a) **Age and gender.** HRV has been found to be dependent on the level of exercise, age and gender of the person. Numbers of studies have found the relationship between health and other parameters with HRV, such as it has been seen that physically active young people have higher HRV [10–11]. Studies on neonatals by Nagy et al. [12] have observed gender and alertness based differences; that the alert new born have lower HR variation in the boys than in the case of girls. Bonnemeir et al. [13] reported that the HRV decreased for healthy subjects from 20 to 70 years and the variation is more in the case of female than men.

b) **Hypertension.** It has been reported that the structural and functional alterations of the cardiovascular system are frequently present in the individuals with hypertension and this may increase their cardiovascular risk [10]. ECG of left ventricular hypertrophy (LVH) and strain are associated with increased morbidity and mortality and this is also observed in HRV as it significantly decreases in patients with LVH secondary to hypertension.

c) **Drugs and medications.** Heart rate variability has been shown to be significantly influenced by various groups of drugs and it is important to consider the influence of medication when interpreting HRV. Research studies [14–15] have reported on the effects of beta-blockers and calcium channel blockers on the heart rate variability in patients with post-infarction and hypertension.

4.2.1 Computing heart rate variability

Heat rate variability has been analysed by various techniques [10]. Some of these techniques are listed below:

a) **Time domain analysis.** Beat-to-beat indices are the important feature extracted using the time domain analysis [10]. It measures the continuous changes of the heart rate and can also be used to identify the rate of change to the HRV. It lends itself to a number of clinical and other applications and is suitable for real-time monitoring. However, it is limited to the study of immediate changes only and thus not suitable for health monitoring applications.

b) **Non-linear analysis.** Recent developments in the theory of nonlinear dynamics have provided better techniques for analysing the signals generated from nonlinear living systems [16–17]. It is now generally recognized that these nonlinear techniques are able to describe the processes generated by biological systems in a more effective way. There are several non-linear analytic algorithms that have been used for HRV analysis; detrended fluctuation analysis, multiscale entropy and fractal analysis [10].

   The nonlinear dynamical techniques are based on the fractal geometry and concept of chaos theory. These techniques have been applied to a number of biological and medical applications with some major applications in the diagnosis of cardiac disorders. The theory of chaos has been used to detect some cardiac arrhythmias [18].

   There are a number of methods to measure the nonlinear parameters of changing cardiac activity. Fractal dimension and entropy of the heartbeat have been used for characterising the ECG to identify pathological conditions. Methods based on chaos theory have been applied for predicting the onset of events such as ventricular tachycardia for detecting congestive heart failure situations using HRV signals [19–20].

c) **Poincaré geometry.** The Poincaré plot is a technique taken from nonlinear dynamics making it easy to plot heartbeat data and allows visualisation of the heart rate variability. It is a plot of the current RR interval with the interval in the previous beat and displays the nature of RR interval fluctuations [10]. Poincaré plot analysis is an emerging quantitative-visual technique, whereby the shape of the plot is categorized into functional classes that indicate the degree of the heart failure in a subject. While a circular shape indicates steady interval, the direction of the ellipse indicates the direction of the change, and fluctuating shape indicates chaotic signal.

4.3 Fractal Properties of ECG

Studies have shown that the physiological signals generated by complex self-regulating systems under healthy conditions may have a fractal temporal structure [4,10]. Fractal geometry based methods have been successfully used in cardiac signal analysis [4]. It has been reported that ECG signals are well modeled as self-affined fractal sets and their properties can be characterised by using fractal dimension [4,6,21]. The fractal nature of the ECG signal can be attributable to the self-similar pattern of the rhythmic cardiac activity. This change in the rhythmic pattern can be measured using the fractal dimension as the self-similar structure will change.

Fractal geometry is a more accurate method to model the natural phenomena compared with calculus based methods. This is because fractal geometry accepts that small differences in the initial conditions can lead to large differences in the later states, and simple phenomena when repeated over multiple scales, can result in very complex and highly functional systems. This understanding has now made fractal concepts usable for a wide spectrum of fields, starting from image analysis, texture modelling, and market analysis to the study of genome sequences [4,6,10,22]. Meesmann et al. [23] demonstrate the self-affine nature of the human heart rate by means of counting the heart beats. Based on the power law, the number of heart beats in a given time interval scales with the length of time [6,23] and modelled as fractal.

Recent studies have shown that the ECG signal can be modelled as multifractal signals [6,9]. In order to extract additional information about the singular properties of such signals, Goldeber et al. [6] reported the use of the wavelet modulus maxima method to physiologic time series. They have reported that the heart rate time series of healthy humans act as multifractal signals [6,9] and this type of complex variability cannot be attributable only to physical activity. They have also reported that the ECG signal from patients with severe heart failure show a breakdown of multifractal scaling. The detection of multifractal scaling in heart rate dynamics have become one of the most important parameters in analysing the pathological ECG signals.

The structural complexity of the ECG signal has been used to describe the regulatory complexity and physiological interconnections [10]. In order to measure this structural complexity, one of the complexity measurements used in the ECG analysis is Approximate Entropy (ApEn) introduced by Steven Pincus [24] in 1991. ApEn is a statistic quantifying regularity and complexity in a wide variety of relatively short (greater than 100 points) and noisy time series data [24]. The Lempel-Ziv (LZ) complexity algorithm is another measure for analysing the complexity of time series [25]. This measure is associated with the number of distinct patterns within the signal and their occurrence rate along a specified series similar to box counting dimension [25–26].

4.4 An Example

In order to better understand the fractal properties of the ECG signal and changes to these properties with disease, an example is presented. This example shows the difference in the fractal dimension of the ECG signal recorded from case and control; case being a person diagnosed with severe Cardiac Autonomic neuropathy, while control is a normal healthy person. Fractal dimension was computed using Box-counting method and Higuchi Algorithm. Figure 4.2 shows the fractal dimension plot for both normal and severe CAN ECG signal. From this figure, it is observed that there is a reduction in fractal dimension for severe CAN when compared with normal ECG signal which denotes the change in the self-similar structure due to the pathological condition.

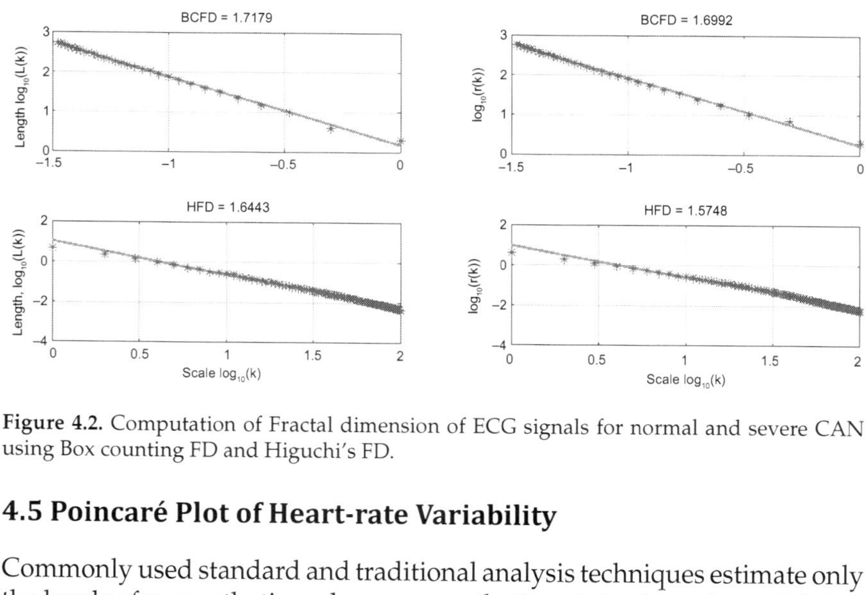

Figure 4.2. Computation of Fractal dimension of ECG signals for normal and severe CAN using Box counting FD and Higuchi's FD.

4.5 Poincaré Plot of Heart-rate Variability

Commonly used standard and traditional analysis techniques estimate only the levels of sympathetic and parasympathetic activity from the variability in the RR intervals. Recently, research studies have investigated [27–29] and reported on two specific HRV analysis techniques [30].

a) Frequency domain - spectral analysis of RR intervals.
b) Poincaré plot of RR interval analysis.

Poincaré plot is a geometrical representation of a time series (RR Interval) in a Cartesian plane. Poincaré HRV plot is a graph in which each RR interval is plotted against next RR interval. Points of the plot are duplets of the values of the time series and the distance (in number of values) between values of each duplet is the lag of the plot. Poincaré plot displays the correlation between consecutive samples in a graphical manner [30].

Qualitative analysis technique, a visual analysis of Poincaré plot is used to distinguish patients with advanced heart failure from healthy individuals. The following patterns are visible for analysing the changes in the RR interval [30,31].

a) *Comet*: Represents the lengthening of RR intervals which indicates increased beat-to-beat variability as well as overall range for healthy subjects [30–31].
b) *Torpedo*: indicates that the change between consecutive RR intervals is minimal [30–31].
c) *Fan*: shows a small increase in RR interval length which is associated with greater dispersion in consecutive RR intervals [30–31].
d) *Complex*: This pattern shows stepwise change in RR intervals which relates to the loss of graded relationship between successive RR intervals and is linked to nonlinear behaviour [30–31].

Ellipse-fitting technique has been mostly used to characterize the Poincaré plot [30]. A sample Poincaré plot is shown in Fig. 4.3.

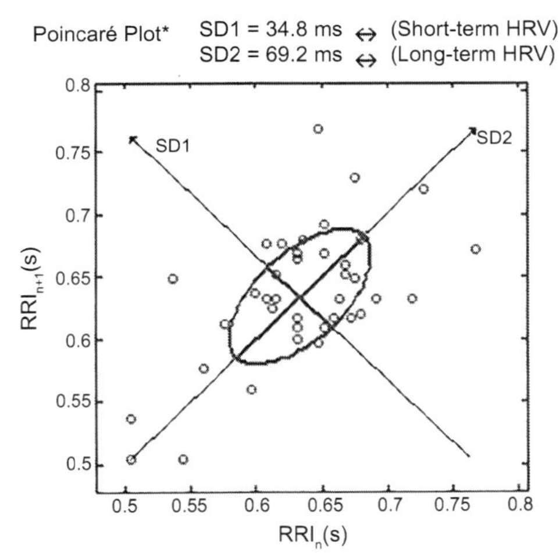

Figure 4.3. An example of Poincaré plot of RR interval recorded from a healthy person.

From the Fig. 4.3, the following parameters are computed [30]:

a) The major axis of the fitted ellipse is aligned with the line of identity. This identity line passes through the origin with slope 45°, and expressed as

$$RR_n = RR_{n+1}$$

b) The minor axis is perpendicular to the line of identity which has a slope of 135° and passes through centroid of the plot and expressed as

$$RR_n + RR_{n+1} = 2\,\overline{rr}$$

where rr represents the RR interval series used in the Poincaré plot and $\overline{rr}$ represents the mean value of RR interval series. In the ellipse-fitting technique, the dispersion of the points along minor axis measures the width of the plot, whereas the dispersion of the points along the major axis measures the length of the plot.

The distance of ith point of the plot, from major axis [30] can be expressed as

$$D_{i(min)} = \frac{RR_i - RR_{i+1}}{\sqrt{2}}$$

and from minor axis [30] can be expressed as

$$D_{i(maj)} = \frac{RR_i + RR_{i+1} - 2\overline{rr}}{\sqrt{2}}$$

Based on these distances, the short term (SD1) of the Poincaré plot are denoted [41] as:

$$SD1 = \sqrt{\frac{1}{N-1}\sum_{i=1}^{N-1} D_{i(min)^2}}$$

and long term variability (SD2) of the Poincaré plot is denoted as

$$SD2 = \sqrt{\frac{1}{N-1}\sum_{i=1}^{N-1} D_{i(maj)^2}}$$

where N is the number of RR intervals in the plotted HRV signal.

4.6 Application—ECG and Heart Rate Variability

Certain cardiac diseases, stress, physical exercises and other pathologic states have been known to show effect on HRV [32–33]. Various research works, for example, have suggested a relationship between negative emotions and reduced HRV [34]. Tzaneva et al. [35] have investigated, based on HRV, the effect of 135 min exposure to noise with intensity—95 dB(A) in three experimental sessions: before, at the onset, and at the end of the noise exposure. According to their results, the comparison of HRV parameters before and on the onset of the experiment showed tendency for significant decrease of the cardio-intervals variability parameters such as standard deviation, sum of positive differences between successive cardio intervals, total wave energy in the cardio-tachogram and mean difference between successive cardio intervals. In this section, we are reporting our work which has been published [32].

a) **Materials.** Disposable Ag/AgCl electrodes (AMBU Blue Sensors from MEDICOTEST) were used to record the ECG using Lead II configuration. Before the recording of ECG commenced, the skin was prepared for the electrodes using light aberration paper and alcohol swipes. After cleaning the skin, the electrodes were affixed at the fourth intercostal space on the right and left side of the sternum on the chest of the subject. All signals were recorded using AMLAB bio-amplifier with flexible real time signal conditioning.

   Towards the aim of identifying the suitability of HRV to measure exercise induced responses, non-dynamic controlled experiments were conducted where subjects performed a simple physical exercise. Five volunteers participated in the experiment. All the participants were males and the average age of the participants was 29 yrs. After recording five minutes of ECG at rest, subjects were asked to perform a physical task (walking up and down five flights of stairs in the University building). The average time required to perform the task for each subject was 5 minutes. A five minute ECG recording was performed immediately after the exercise.

b) **Methods.** Data processing and extraction of parameters was performed off-line using MATLAB software. The first step of the analysis involves extracting RR interval from the raw ECG signal. In order to achieve this, accurate identification and estimation of the inter-beat intervals based on the R-wave peak as the reference point was required. This was accomplished by implementing a signal peak (R-wave peak) detection algorithm.

HRV was then obtained from the estimated RR intervals and the following analyses were performed:

- *time domain,*
- *frequency domain,*
- *Poincaré analysis,*
- *Fractal dimension.*

Time domain analysis

Tables 4.1 and 4.2 show the descriptive statistics of the HRV and RR extracted from the ECG signal before and after exercise for a subject. As expected, after the exercise, the mean HR increased in all subjects. The terms used in the tables as described by Task Force of the European Society of Cardiology and the North American Society of Pacing Electrophysiology [5] are listed below:

- *NN* – Normal to Normal intervals – all intervals between adjacent QRS complexes.
- *RMSSD* – the square root of the mean squared differences of the successive NN intervals NN50 – the number of interval differences of successive NN intervals greater than 50 ms.
- *pNN50* – the proportion derived by dividing NN50 by total number of NN intervals.
- *RR triangular index* – the number of all RR intervals divided by the maximum of the density distribution.
- *Triangular interpozation of NN interval histogram (TINN)* – Baseline width of the minimum square difference triangular interpolation of the highest peak of the histogram of all NN intervals
- *VLF* – Power in very low frequency range
- *LF* – Power in low frequency range
- *HF* – Power in high frequency range

Table 4.1. Descriptive statistics of the HRV and RR extracted from the ECG signal before the exercise for a subject.

Time Domain Statistics		
Variable	Units	Value
Statistical Measures		
Mean RR*	(s)	0.775
STD	(s)	0.082
Mean HR*	(1/min)	78.57
STD	(1/min)	11.12
RMSSD	(ms)	94.6
NN50	(count)	45
pNN50	(%)	28.8
Geometric Measures		
RR triangular index		0.088
TINN	(ms)	400.0

Distributions*

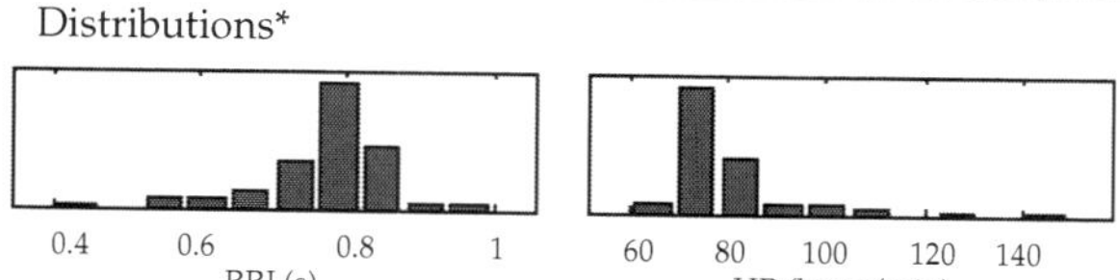

* Calculated from the non-detrended selected RRI signal.

Table 4.2. Descriptive statistics of the HRV and RR extracted from the ECG signal after the exercise for a subject.

Time Domain Statistics		
Variable	Units	Value
Statistical Measures		
Mean RR*	(s)	0.740
STD	(s)	0.102
Mean HR*	(1/min)	83.01
STD	(1/min)	13.68
RMSSD	(ms)	130.8
NN50	(count)	83
pNN50	(%)	46.4
Geometric Measures		
RR triangular		0.127
index	(ms)	450.0
TINN		

Distributions*

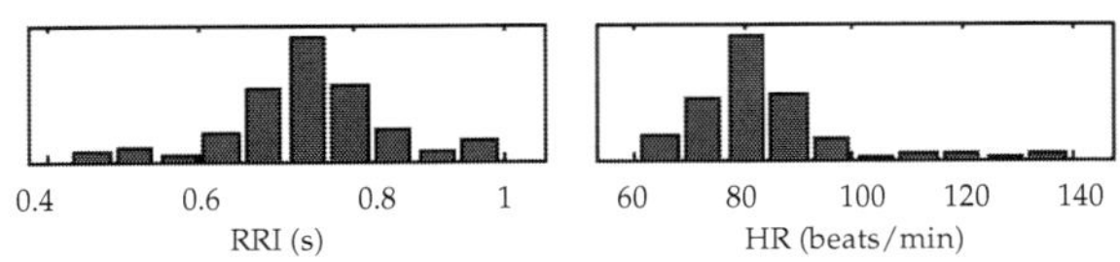

* Calculated from the non-detrended selected RRI signal.

Frequency domain analysis

The frequency domain analyses with the HRV analysis for LF, HF and LF/HF of the data extracted before and after exercise have been shown in Fig. 4.4 and Fig. 4.5. The LF% and HF% were obtained from the power spectra of the HRV data. At rest, from the autoregressive (AR) spectrum, it can be seen that LF/HF ratio is higher than the ratio observed after the exercise.

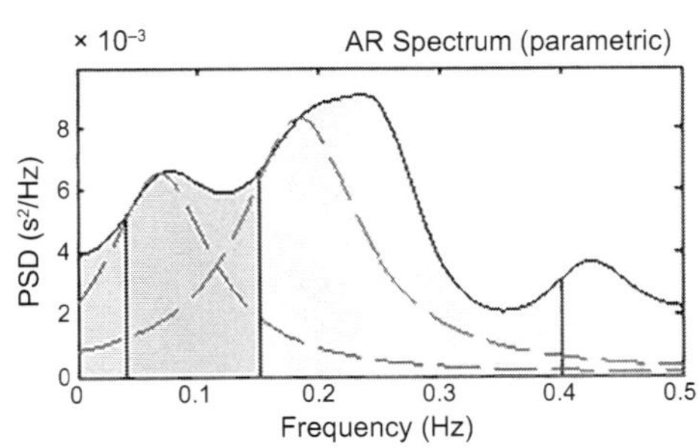

Frequency Band	Peak (Hz)	Power (ms²)	Power (%)	Power (n.u.)
VLF	0.0000	0	0.0	
LF	0.0684	679	47.1	22.9
HF	0.1836	763	52.9	25.8
LF/HF			0.890	

Figure 4.4. AR Spectrum Analysis before exercise.

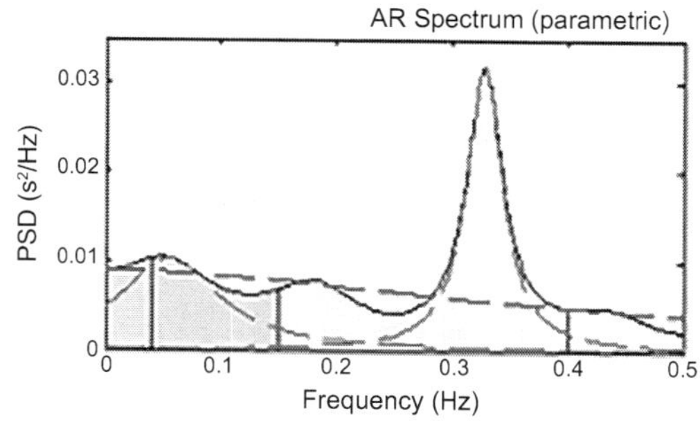

Frequency Band	Peak (Hz)	Power (ms^2)	Power (%)	Power (n.u.)
VLF	0.0000	0	0.0	
LF	0.0566	1091	38.3	23.4
HF	0.3281	1759	61.7	37.7
LF/HF			0.620	

Figure 4.5. AR Spectrum Analysis after exercise.

The resultant LF/HF power appears to be the most sensitive measure that indicates the change in cardiac activity in response to exercise. The HF component of the HRV has been reported to reflect activity of the vagal parasympathetic activity mostly modulated by respiration while LF component is influenced by both sympathetic and parasympathetic activities. The LF/HF ratio reflects the sympathetic-vagal balance and describes ANS [37]. Literature review, however, indicates that there exist disagreements with respect to the LF component. Some studies suggest that LF is a quantitative marker for sympathetic modulations. Other studies view LF as reflecting both sympathetic and vagal activity while the LF/HF ratio being considered by some investigators to mirror vagal-sympathetic balance [5]. As discussed earlier, the heart rate is regulated by increased sympathetic activity and reduced parasympathetic activity, causing the heart rate to rise. The relative roles of the two activities depend on the intensity of the exercise [37].

Poincaré analysis

Figures 4.6 and 4.7 display the Poincaré plots before and after the exercise respectively. SD1 (short term with short diameter) and SD2 (long term with long diameter) of the Poincaré plot are used to display the statistical distribution of the signal. SD1 describes the fast beat-to-beat HRV. It represents statistical dispersion perpendicular to the straight line of identity. Similarly, SD2 measures dispersion along the straight line describing the slower components of HRV. In all experimental cases, SD1 and SD2 increased after the exercise for both short and long term HRV showing the changes due to the physical activity.

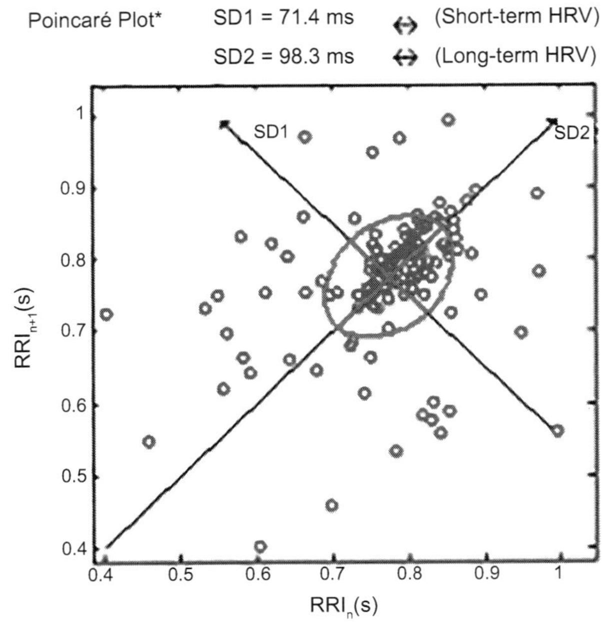

Figure 4.6. Poincaré plot of HRV before exercise.

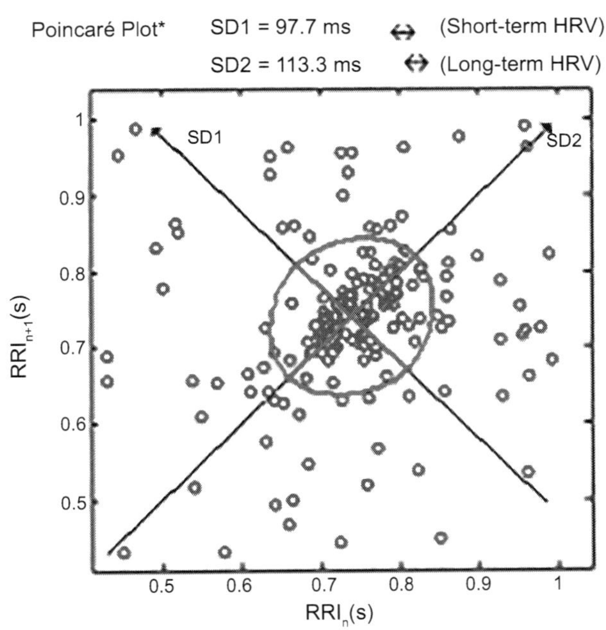

Figure 4.7. Poincaré plot of HRV after exercise.

Fractal dimension

In order to observe the fractal properties of the ECG signal recorded before and after physical exercise, fractal dimension of the ECG signal was computed using Box counting method. Figure 4.8 shows the mean fractal dimension before and after physical exercise. From the figure, we can observe that there is a decrease in the mean fractal dimension after the physical exercise. The results are in line with the study [38] reporting changes in FD due to the physical exercise.

It has been reported that the fractal dimension of HRV can provide a measure for the complexity of the autonomic centers [52] in the body and will reflect the number of dominant inputs that have impact on these autonomic centers and their interactions. Study by Nakamura et al. [38] has shown that fractal dimension of HRV decreased gradually during mild to high exercise intensity.

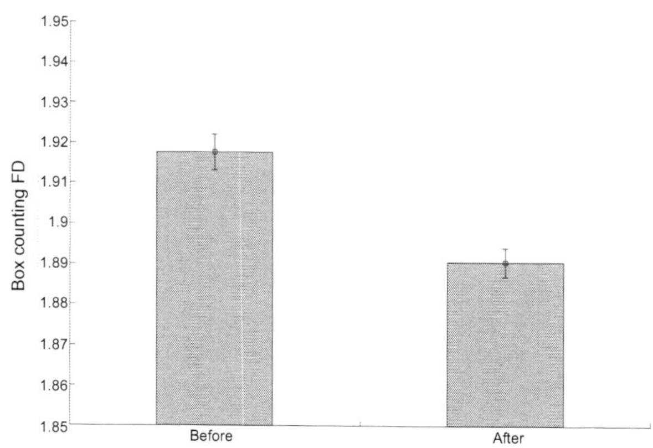

Figure 4.8. Mean (SD) fractal dimension of ECG recorded before and after physical exercise.

4.7 Summary

Monitoring and characterisation of heart rate activity is essential for diagnosing cardiovascular disease. Electrocardiogram (ECG) signal is the non-invasive recording of the cardiac activity and has been analysed to understanding and monitoring the heart. While there are a number of techniques and methods that have been developed to analyse the ECG signal, it is now acknowledged that cardiac activity is not regular and periodic but chaotic which is best described by fractal geometry and non-linear analysis. This chapter has discussed the fractal characteristics of the ECG signal, described different techniques to compute its fractal dimension and examine the differences between healthy and CAN patient. Heart-rate variability has been currently used as an important tool to measure the changes in the characteristics of the heart. Poincaré plot is one of the techniques that facilitate the analysis of the long range and short range variability of the heart-beat. Visual analysis of the Poincaré plot provides a method for examination of the signal and provides better understanding of the characteristics of the HRV signal. Current research considers the HRV signal as a multi-fractal signal where it is modelled with multiple indices not as singularity characteristics.

References

1. CDC/National Center for Health Statistics, http://www.cdc.gov/nchs/fastats/leading-causes-of-death.htm. Access Date: 18 March 2016.
2. Anatomy of the Heart, http://www.qcg.com.au/anatomy_of_heart.html. Access Date: 18 March 2016.
3. What is ECG and how does it work?, https://imotions.com/blog/what-is-ecg/ Access Date: 18 March 2016.
4. Turcott, R.G. and M.C. Teich. 1996. Fractal character of the electrocardiogram: Distinguishing heart-failure and normal patients. Annals of Biomedical Engineering, 24(2): 269–293.
5. Heart Rate Variability: Standards of Measurement, Physiological Interpretation, and Clinical Use. Task Force of the European Society of Cardiology the North American Society of Pacing Electrophysiology Circulation. 1996, 93: 1043–1065.
6. Goldberger, A.L., L.A.N. Amaral, J.M. Hausdorff, P.Ch. Ivanov, C.-K. Peng and H.E. Stanley. 2002. Fractal dynamics in physiology: Alterations with disease and aging. Proceedings of National Academy of Science of the USA 2002, 99(suppl 1): 2466–2472.
7. Stanley, H.E., L.A.N. Amaral., A.L. Goldberger, S. Havlin, P.Ch. Ivanov and C.-K. Peng. 1999. Physica A, 270: 309–324.
8. Ivanov, P.Ch., L.A.N. Amaral, A.L. Goldberger, S. Havlin, M.G. Rosenblum, H.E. Stanley and Z.R. Struzik. 2001. Chaos, 11: 641–652.
9. Ivanov, P.Ch., L.A.N. Amaral, A.L. Goldberger, S. Havlin, M.G. Rosenblum, Z. Struzik and H.E. Stanley. 1999. Multifractality in Human Heartbeat Dynamics, Nature (London), 399: 461–465.
10. Rajendra Acharya, U., K. Paul Joseph, N. Kannathal, Choo Min Lim and Jasjit S. Suri. 2006. Heart rate variability: a review. Medical and Biological Engineering and Computing, December 2006, 44(12): 1031–1051.
11. Ryan, S.M., A.L. Goldberger, S.M. Pincus, J. Mietus and L.A. Lipsitz. 1994. Gender- and age-related differences in heart rate: are women more complex than men. J. Am. Coll. Cardiol., 24(7): 1700–1707.
12. Nagy, E., H. Orvos, G. Bardos and P. Molnar. 2000. Gender related heart rate differences in human neonates. Pediatr. Res., 47(6): 778–780.
13. Bonnemeier, H., U.K.H. Wiegand, A. Brandes, N. Kluge, H.A. Katus, G. Richardt and J. Potratz. 2003. Circadian profile of cardiac autonomic nervous modulation in healthy subjects: Differing effects of aging and gender on heart rate variability. J. Cardiovasc. Electrophysiol., 14: 8791–799.
14. Coumel, P., J.S. Hermida, B.Wennerblom, A. Leenhardt, P. Maison-Blanche and B. Cauchemez. 1991. Heart rate variability in left ventricular hypertrophy and heart failure, and the effects of beta-blockade. Eur. Heart J., 12: 412–422.
15. Guzzetti, S., E. Piccaluga, R. Casati, S. Cerutti, F. Lombardi, M. Pagani and A. Malliani. 1988. Sympathetic predominance inessential hypertension: a study employing spectral analysis of heart rate variability. J. Hypertens., 6: 711–717.
16. West, B.J. 2000. Fractal physiology and medicine: studies of nonlinear phenomena in life. Science 1. World Scientific, Singapore.

17. Akay, M. 2001. Nonlinear biomedical signal processing dynamic analysis and modeling, vol. II. IEEE Press, New York.
18. Kaplan, D.K. and J.R. Cohen. 1991. Searching for chaos in fibrillation. Ann. NY Acad. Sci., 367–374.
19. Cohen, M.E., D.L. Hudson and P.C. Deedwania. 1996. Applying continuous chaotic modeling to cardiac signal analysis. IEEE Eng. Med. Biol., 15: 97–102.
20. Rajendra Acharya, U., N. Kannathal and S.M. Krishnan. 2004. Comprehensive analysis of cardiac health using heart rate signals. Physiol. Meas., 25: 1139–1151.
21. Meyer, M. and O. Stiedl. 2003. Self-affine fractal variability of human heartbeat interval dynamics in health and disease. European Journal of Applied Physiology, 90(3): 305–316.
22. Bassingthwaighte, J., L. Liebovitch and B. West. 1994. Fractal Physiology, Oxford University Press, New York.
23. Meesmann, M., D.R. Jan boese, P.K. Chialvo, R.B. Wolfgang, P. Werners, G. Ferdinand and K. Klaus-dietrich. 1993. Demonstration of 1/F fluctuations and white noise in the human heart rate by the variance-time-curve: Implications for self-similarity. Fractals, 01: 03, 312–320.
24. Pincus, S.M. 1991. Approximate entropy as a measure of system complexity. Proc. Natl. Acad. Sci. USA, 88: 2297–2301.
25. Aboy, M., R. Hornero, D. Abásolo and D. Alvarez. 2006. Interpretation of the Lempel-Ziv complexity measure in the context of biomedical signal analysis. IEEE Trans. Biomed. Eng., 53(11): 2282–2288, Nov. 2006.
26. D. Abásolo and H.F. Jelinek. 2012. Lempel-Ziv complexity dynamics in early detection of cardiac autonomic neuropathy in diabetes.
27. Nakao, M., M. Norimatsu, Y. Mizutani and M. Yamamoto. 1997. Spectral distortion properties of the integral pulse fre quency modulation model. IEEE Trans. Biomed. Eng., 44(5): 419–426.
28. Mohn, R.K. 1976. Suggestions for the harmonic analysis of point process data. Comput. Biomed. Res., 9: 521–530.
29. Hyndman, R.D. and R.K. Mohn. 1975. A model of the cardiac pacemaker and its use in decoding the information content of cardiac intervals. Automedica, 1: 239–252.
30. Khandoker, A.H., C. Karmakar, M. Brennan, M. Palaniswami and A. Voss. 2013. Poincaré plot methods for heart rate variability analysis. New York: Springer.
31. Woo, M.A., W.G. Stevenson, D.K. Moser, R.B. Trelease and R.M. Harper. 1992. Patterns of beat-to-beat heart rate variability in advanced heart failure. Am. Heart J., 123(3): 704–710.
32. Alemu, M., S.P. Arjunan and D.K. Kumar. 2011. Observing exercise induced heart rate variability response. Biosignals and Biorobotics Conference (BRC), 2011 ISSNIP, Vitoria, pp. 1–6.
33. Rajendra Acharya, U., N. Kannathal, O.W. Sing, L.Y. Ping and T.L. Chua. 2004. Heart rate analysis in normal subjects of various age groups. Biomed. Eng. Online, vol. 3, no. 1.
34. Kawachi, I., D. Sparrow, P.S. Vokonas and S.T. Weiss. 1995. Decreased heart rate variability in men with phobic anxiety. Am. J. Cardiol., 75: 882–885.
35. Tzaneva, L., S. Danev and R. Nikolova. 2001. Investigation of noise exposure effect on heart rate variability parameters. Cent. Eur. J. Public Health, 9(3): 130–132.
36. Niskanen, J., M.P. Tarvainen, O. Perttu, Ranta-aho and P.A. Karjalainen. 2004. Software for advanced HRV analysis. Computer Methods and Programs in Biomedicine, 76(1): 73–81.
37. Bernardi, L., F. Salvucci, R. Suardi, P.L. Solda, S. Perlini, C. Falcone and L. Ricciardi. 1990. Evidence for an intrinsic mechanism regulating heart rate variability in the transplanted and intact heart during submaximal exercise. Cardiovasc. Res., 24: 969–981.
38. Nakamura, Y., Y. Yamamoto and I. Muraoka. 1993. Autonomic control of heart rate during physical exercise and fractal dimension of heart rate variability. Journal of Applied Physiology Feb. 1993, 74(2): 875–881.

CHAPTER 5

Fractals Analysis of Surface Electromyogram

ABSTRACT

Fractal theory has been used in the analysis of physiological time series due to its complexity. The nonlinearity of physiological systems may have relevance for modelling complicated surface electromyogram (sEMG) for example, low-level movements in which interactions and cross-talk occur over a wide range of temporal and spatial scales. Fractal theory based analysis is one of the most promising new approaches for extracting such hidden information from physiological time series signal like sEMG, which can provide information regarding the characteristic temporal scales and the adaptability of muscle activity response. This chapter investigates the use of fractal theory for analysis of EMG signal for applications in rehabilitation and age-related changes in the muscle properties and contraction.

5.1 Introduction

Biosignals such as sEMG are a result of the summation of similar motor unit action potentials (MUAP) that travel through tissues and undergo spectral and magnitude compression. Visual inspection of the signal shows that it has a noisy appearance with an inherent randomness. Its properties such as magnitude that have often been considered to assess the muscle activity, however are highly variable because of a range of intrinsic and extrinsic factors.

The randomness associated with the signal is an important observation of the signal, and useful for assessment of the muscle activity. One aspect of this is associated with the similarity of the different MUAP, and the temporal burst within burst behaviour of the signal. The signal demonstrates the property that patterns observed at one sampling rate are statistically similar to patterns observed at lower sampling rates thereby suggesting the self-similar characteristics of sEMG [1].

Objects or signal patterns that exhibit self-similarity have fractional dimension and are referred to as Fractals. These objects or patterns on any level of magnification will yield a structure that resembles the larger structure in complexity [2]. The measured property of the fractal process is scale dependant and has self-similar variations in different time scales. Fractal dimension of a process measures its complexity, spatial extent or its space filling capacity and is related to shape and dimensionality of the process [3]. The concept of fractal has been applied effectively to physiological process which have self-similarity over multiple scales in time and its spectrum is not narrow band. Signals representing such process have a high degree of visual complexity [4]. Electromyogram is a signal exhibiting the characteristics of self-similarity where the shape of the motor unit action potential has been observed over multiple scales.

Investigations have revealed that sEMG can be considered to have fractal properties [3] and that these are suitable to characterise the EMG signal based on different muscle activations [4,5]. It has also been observed that fractal dimension (FD) of sEMG are suitable for estimating muscle activity parameters from the recordings [1,6,7]. Some of the recent and past studies that have reported the fractal analysis of sEMG include works by Hu et al. [5], Anmuth et al. [1], Arjunan et al. [6–8], Gitter et al. [3], Gupta et al. [4], and Beretta Picolloi et al. [9].

Hu et al. [5] distinguished two different kinds of limb actions—Forearm supination and forearm pronation using FD of sEMG signals. Anmuth et al. [1] observed the relationship between muscle activity and fractal dimension of the surface EMG signal and determined that FD of sEMG was linearly related to the fraction of maximum voluntary contraction. They also observed a linear relationship between the fractal dimension and the flexion-extension speeds and load.

Arjunan et al. [6–8] reported the use of fractal features of EMG to identify various patterns of finger and wrist movements for the applications in the prosthetic control. They have also reported the use of fractal features to investigate the age-related neuromuscular changes [10].

Gitter et al. [3] determined that fractal dimension can be used to quantify the complexity of motor unit recruitment patterns. They also demonstrated that the fractal dimension of sEMG signal is correlated with muscle force. Gupta et al. [4] found that the FD of EMG could be used to characterize the flexion-extension of the arm.

Fractal dimension represents the scale invariant non-linear property of the source of the signal and is an index for describing the irregularity of a time series. FD is the property of the system or source of the signal (reference) and in the case of sEMG, it is the property of the muscle. It should be a measure of the muscle complexity and not a measure of the level of muscle activity. Research study by [4,5,11] has attributed change in FD to change in level of muscle contraction during high level muscle activity.

Study by Basmajian and De Luca [12] have indicated that for low level of isometric muscle contraction, there is no change in the size of the muscle while there is measurable change in the muscle dimension during higher levels of muscle contractions and during non-isometric contraction.

5.2 Surface Electromyogram (sEMG)

The fundamental structure of the muscle consists of muscle fibres or cells. These are grouped in units such that all the fibres in one unit are activated by one neurone. Each of these units is called the motor unit or MU, and the fibres of each MU are activated nearly simultaneously. When the muscle fibre is activated, an electrical pulse travels across the length of the fibre and the summation of the electrical potentials from all the fibres in a single MU are called motor unit action potential or MUAP.

Electromyogram is the electrical signal associated with the contraction of the skeletal muscles. It is a result of the electrical potential generated by the summation of all the MUAP. As the electrical pulses travel through the tissues, there is significant attenuation and spectral compression, and hence the recording is influenced by the location of the electrodes with reference to the active muscles.

Electromyogram may be recorded using needle electrodes or surface electrodes. The needle electrodes are fine metallic needles that are inserted inside the muscle and record the electrical activity from close proximity to the muscle fibres. Because of high rate of attenuation of the signal, needle EMG is specific to the active muscle fibres that are alongside the electrode.

Surface electromyography (sEMG) is the recording of the muscle's electrical activity from the surface of the skin. This activity is gross, and is the summation of all the MUAP in the proximity. While needle EMG tends to have a series of sharp spikes, sEMG is a lower frequency signal with noisy appearance. sEMG is used for the diagnosis of neuro-muscular disorder and for rehabilitation. It is also used for device control applications where the signal is used for controlling devices such as prosthetic devices, robots, and human-machine interface.

The advantage of sEMG is due to its non-invasive recording technique and it provides a safe and easy recording method. The underlying mechanism of sEMG is very complex [12] because there are number of factors such as neuron discharge rates, motor unit recruitment and the anatomy of the muscles and surrounding tissues that contribute to the recording. The signal is dependent on large number of intrinsic and external factors and appears very noisy. Thus, obtaining suitable information that is related to specific muscle properties is often a challenge. In this chapter, the basic concepts of generation of sEMG signals and its application will be described.

5.2.1 Principles of sEMG

sEMG signal is the electrical activity induced because of the electrical activity of the active muscle fibers during a contraction. The active muscle fibres are in the depolarised state and behave like signal sources. These are separated from the recording electrodes by tissues, which act as spatial low-pass filters on the (spatial) potential distribution [12]. The signal recording is a gross representation of the muscle activity, and is influenced by muscle size, number of active motor units, rate of activation of the motor units and the functional state of muscle fibres (fatigue) [12]. sEMG signal is generated by the electrical activity of the muscle fibers active during a contraction. The signal sources located at the depolarized zones of the muscle fibers are separated from the recording electrodes by biological tissues, which act as spatial low-pass filters on the (spatial) potential distribution [12]. It is closely related to the muscle activity, muscle size and a measure of the functional state of muscle fibres [13].

5.2.2 Factors that influence sEMG

The action potentials recorded in sEMG signals are generated by the electrical activities in the muscle. The signal contains information related to muscle contraction and its condition. However, there are number of factors that influence the signal and it is essential to examine these to ensure appropriate interpretation of the signal. These factors are listed below.

- *Level of Contraction*
  Rate of muscle activity
  Number of active motor units
  Size of the active motor units
- *Localised Muscle Fatigue*
- *Body Tissue*
  Skin thickness
  Size of muscle
  Thickness of fat layer
- *The Inter-electrode Distance*
- *The Artefacts and Noises*
  Motion artefacts

Line noise and harmonics

Circuit noise

- *Crosstalk – signals from adjoining muscles*
- *Electrode to skin contact impedance*

Reference contact impedance

Electrode to skin impedance

Electrode gel

From the above, it is evident that there is large number of factors that affect the signal. While some of these can be reduced by carefully controlled experiments, most of these cannot be controlled. Hence it is essential to identify the relationship of the features of the signal with the physiological factors.

5.2.3 Signal features of sEMG

Over the past few decades, numerous features of sEMG have been investigated. These can be largely divided in three categories; *amplitude based, spectrum based*, and *Statistical and chaos based features*:

Amplitude analysis

Amplitude of sEMG is evidently closely associated with the strength of the muscle activity. Prior to the widespread availability of computers, it was also the easiest and most practical to measure. Due to the stochastic nature of the signal, direct measure of the amplitude is not appropriate and one of the commonly used features is Root Mean Square (RMS) of the signal. The squaring of the signal overcomes the biphasic nature of the signal, while windowing captures the strength of the signal over a small time window. There are number of other features that are also used for estimating the strength of the signal, two of which are; *integration* and *rectification*.

Spectral analysis

Spectrum of sEMG is closely associated with rate of muscle activity, conduction velocity and the status of muscle fatigue. Spectrum analysis has number of applications such as identifying the envelope of the signal during cyclic activity, and for muscle fatigue analysis. While median frequency of the signal was the most commonly used measure of the spectrum, recently normalised spectral moments [14–15] are being used as the features. Time frequency analysis and wavelet transforms are also used to identify changes in the spectral properties.

Statistical and chaos based features

Due to the recent advancement in the computational power more features of the signal have been investigated that are neither directly a measure of the amplitude nor the spectrum of the signal. The features based on the statistical properties and chaos theory have been recently reported in the various research studies in analysing the signal. Independent component analysis based methods such as Increase in synchronisation (IIS) Index [16], non-negative matrix factorisation have been reported to analyse the EMG signal which provides significant features of the EMG signal for further analysis.

5.3 Fractal Analysis of sEMG

Rehabilitation process, clinical diagnosis and basic investigations are critically dependent on the ability to record and analyse physiological signals like ECG, EEG and EMG. However, the traditional analyses of these signals have not kept pace with major advances in technology that allow for recording and storage of massive datasets of continuously fluctuating signals. Although these typically complex signals have recently been shown to represent processes that are non-linear and non-stationary in nature, the methods used to analyse these data are often assumed linearity and stationary-like conditions. Such conventional techniques include analysis of means, standard deviations and other features of histograms, along with classical power spectrum analysis.

Recent findings [9,21] show that sEMG signals may contain hidden information that is not extractable with conventional methods of analysis. Such hidden information promises to be of clinical value as well as to relate to basic mechanisms of muscle property and activity function. Fractal theory based analysis is one of the most promising new approaches for extracting such hidden information from physiological time series signal like sEMG, which can provide information regarding the characteristic temporal scales and the adaptability of muscle activity response [3–9,21].

5.3.1 Self-similarity of sEMG

In complex bio signals like sEMG, there exists self-similarity phenomenon, in which there is a small structure (Motor Unit) that statistically resembles the larger structure. The source of EMG is a set of similar action potentials originating from different locations in the muscles. Because of the self-similarity of the action potentials that are the source of the EMG recordings over a range of scales, EMG has fractals properties.

Preliminary analysis was performed to establish the suitability of the use of fractal analysis of sEMG recordings. The recording of sEMG while performing simple contraction was conducted to test the presence of self-similarity. The self-similarity property of sEMG was tested using the following procedure mentioned in the study by Kalden and Ibrahim [22].

- If $y(k)$ be a time series signal, then $y^m(k)$ is the *aggregated process* with non-overlapping blocks of size 'm' such that:

$$y^{(m)}(k) = \frac{1}{m}\sum_{l=0}^{m-1} y(km-l) \tag{5.1}$$

- For the signal or process, y (k) to be self-similar, the variance of the aggregated process decays slowly with m and this self-similarity is measurable by H, that is,

$$Var(y^{(m)}) \approx \boldsymbol{m}^{-\beta} \tag{5.2}$$

with $0 < \beta < 1$ and

$$H = 1 - \beta/2 \tag{5.3}$$

where H expresses the degree of self-similarity; large values indicate stronger self-similarity.

From Fig. 5.1, it is observed that the variance decays slowly and the self-similarity measure, β is less than 1 (H = 0.6), which shows the self-similarity nature of sEMG.

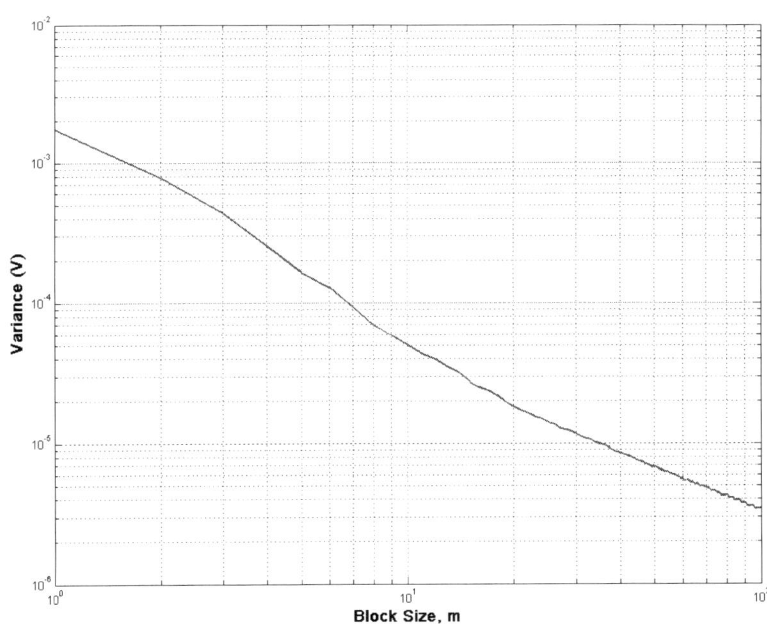

Figure 5.1. Plot of self-similarity measure of the surface electromyogram signal.

5.3.2 Algorithms to compute fractal dimension of sEMG

Fractal dimension (FD) analysis is frequently used in physiological signal processing like sEMG, EEG, and ECG [23–25]. Applications of FD in these physiological signals include two types of approaches [26,27,33]:

Signals in the time domain

This approach estimates the FD directly in the time domain or in the original waveform domain, where the waveform or original signal is considered a geometric figure.

Signals in the phase space domain

Phase space approaches estimate the FD of an attractor in state space domain. Calculating the FD of waveforms is useful for transient detection, with the additional advantage of fast computation. It consists of estimating the dimension of a time-varying signal directly in the time domain, which allows significant reduction in program run-time [26]. Fractal dimension of sEMG is calculated to determine the transients in sEMG that is related to the overall complexity of the muscle properties.

As explained in Chapter 3, Section 3.3, the following prominent methods for computing the FD of a waveform [28–30] have been applied to the analysis of signals, and a variety of engineering systems

- Box counting Method
- Higuchi's Algorithm [28]
- Katz's Algorithm [29]
- Petrosian's Algorithm [30]

Studies by Esteller et al. [31] have shown that Higuchi's algorithm provides the most accurate estimates of the FD. Katz's method was found to be less linear and its calculated FD was exponentially related to the known FD, whereas Petrosian's algorithm was found to be relatively linear and demonstrated the least dynamic range for the estimated FD. Based on this, Higuchi's algorithm was considered for the computation of FD of sEMG in this study.

5.3.3 Fractal features of sEMG

In order to study the fractal characteristics of the sEMG signal, FD was calculated using the Higuchi algorithm [28,31] for non-periodic and irregular time series. This algorithm yields a more accurate and consistent estimation of FD for physiological signals than other algorithms [31].

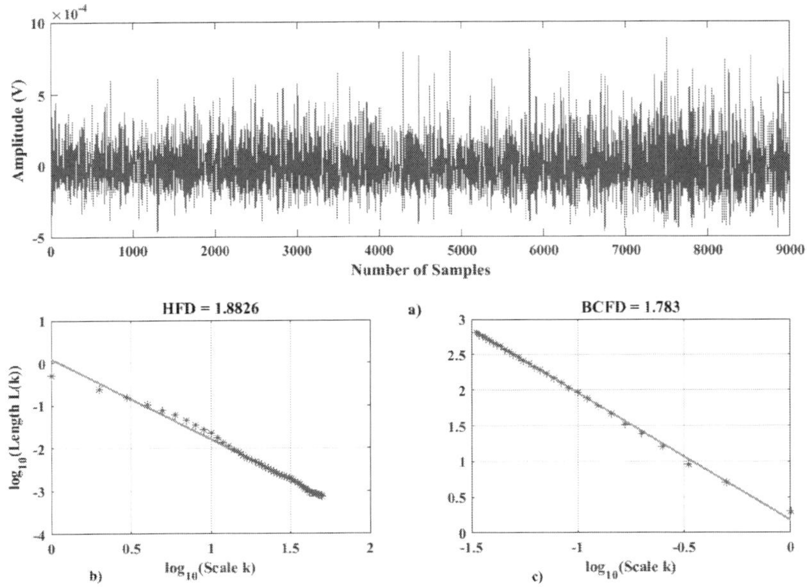

Figure 5.2. Example Plot of computing FD of sEMG (a) Sample sEMG signal (b) Higuchi's FD (c) Box counting FD.

Fractal dimension (FD) of sEMG has been found to be sensitive to magnitude of the muscle and level of force of muscle contraction. FD is the index that describes the irregularity of a time series in place of the power law index. Gitter et al. [3] demonstrated that the fractal characteristics of EMG signal with a dimension is highly correlated with muscle force. Gupta et al. [4] reported that the fractal dimension can be used to characterize the EMG signal. Hu et al. [9] distinguished two different patterns of sEMG signals using fractal dimension. These studies demonstrated that fractal information of sEMG is useful for characterizing the signal and identifying properties of the signal.

Recently Boccia et al. [32] and Beretta-Piccoli [9] have shown the relationship of fractal dimension of sEMG during aging and fatiguing contractions. They have reported that there is a significant correlation between FD and muscle conduction velocity at submaximal muscle contractions which is probably as a result of the motor unit synchronisation. Researchers have studied fractal dimension to characterize normal and pathological signals [23]. Anmuth et al. [1] determined that there was a small change of the fractal dimension of the EMG signal and this was linearly related to the activation of the muscle measured as a fraction of maximum voluntary contraction. They also observed a linear relationship between FD and the speed of flexion – extension, and FD with the load being moved.

Works by Arjunan et al. [6] found that the association of the FD of sEMG with force was not significant. They explained the difference based on the fundamental properties of FD, which measures the complexity of the underlying process and the spatial extent or its space filling capacity and is related to shape and dimensionality of the process [2,3]. Based on this fundamental property of the signal, it is argued that the relationship of FD with muscle contraction could be attributed to the change in muscle shape and size due to contraction, because even during isometric contraction, the length of the muscle shortens and the diameter increases.

The inherent properties of any muscle are its size, shape and number of motor units. The number of active motor units would indicate its complexity. These properties would be different for different muscles and would change when a muscle contracts, gets fatigued or due to relative change in the location of the muscle with electrodes. Based on this, it was hypothesised by Arjunan et al. that FD of sEMG recorded from the forearm would change when different digital muscles are activated corresponding to flexion of different fingers. To test this, Arjunan et al. conducted experiments [7,8] and reported the relationship of FD features with maintained finger flexion.

In order to observe the significance of the fractal features, SEMG signals were recorded from four recording locations in the surface of the forearm as shown in Fig. 5.3. The fractal features – Fractal dimension and Maximum Fractal Length—were computed to observe the change in the complexity and strength in muscle activation during different finger and wrist flexions. Maximum Fractal Length (MFL) was computed as the average fractal length of the signal (over unit time) measured at the smallest scale using Higuchi's algorithm, which has been related to the strength of muscle contraction [8]. MFL has also been defined as the modified Waveform length using Root mean square [35] and expressed in eqn. 5.4 and shown in the Fig. 5.4.

$$\mathrm{MFL} = \log_{10}\left(\sqrt{\sum_{n=1}^{N-1}(x(n+1)-x(n))^2}\right) \tag{5.4}$$

Figure 5.5 shows the plot of the fractal features for four different flexion —F1— All fingers and wrist flexion, F2—Index and Middle finger flexion, F3—Wrist flexion towards little finger, F4—Little and ring finger flexion. Based on the complexity and strength of muscle contraction, the four different finger movements can be differentiated using these features. It is observed that the fractal dimension reflects the change in the complexity in muscles activated to perform particular movement.

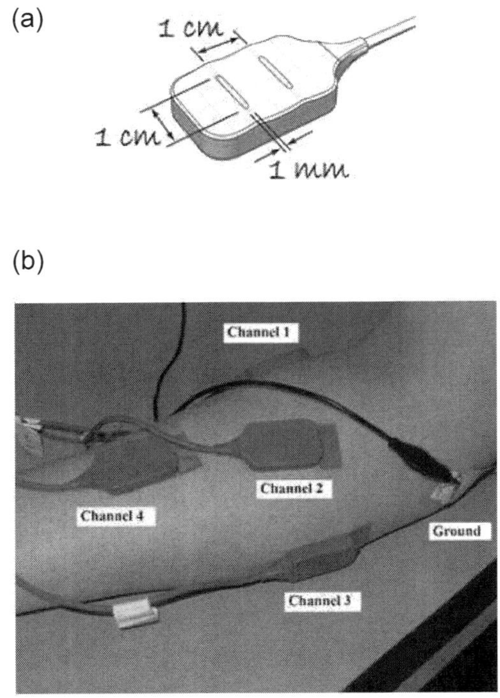

Figure 5.3. Four recording locations in the surface of the forearm.

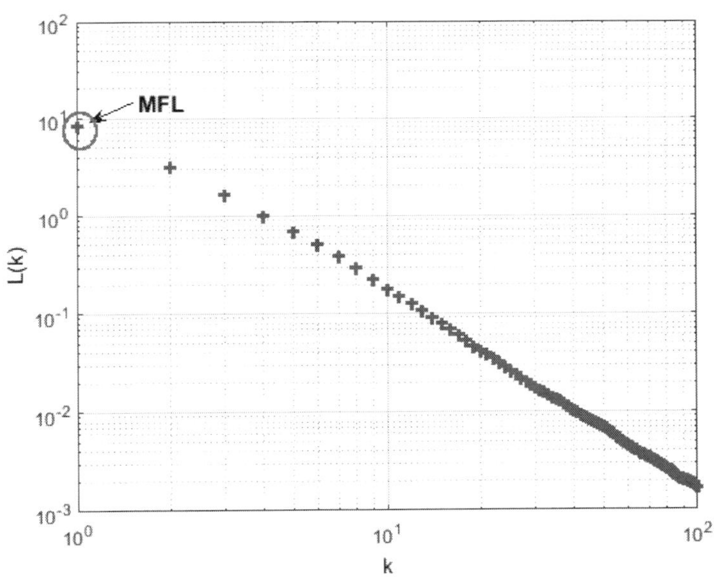

Figure 5.4. Computation of Maximum fractal Length.

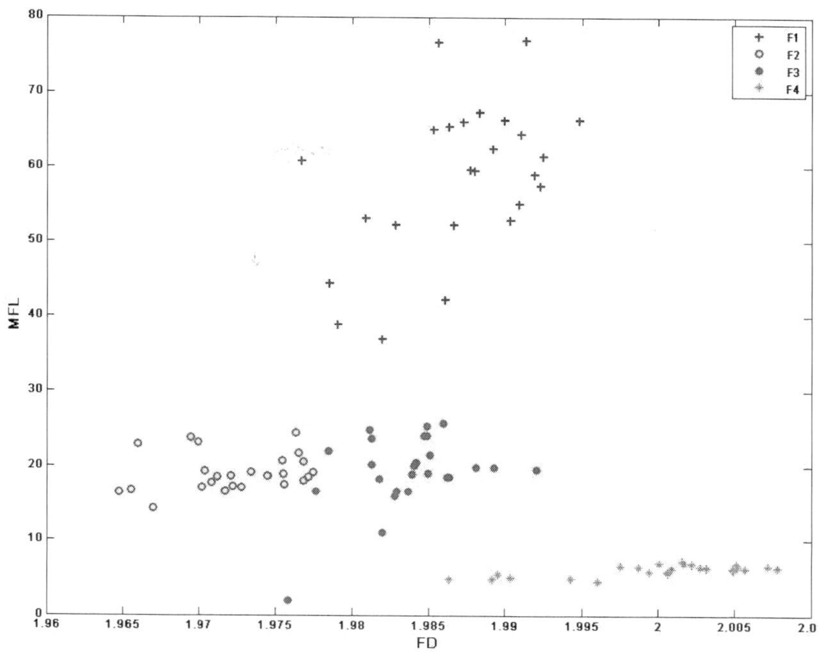

Figure 5.5. Scatter plot of fractal features for four different flexions F1—All fingers and wrist flexion, F2—Index and Middle finger flexion, F3—Wrist flexion towards little finger, F4—Little and ring finger flexion.

5.4 Summary

Clinical diagnosis and basic health investigations are critically dependent on the ability to record and analyse physiological signals like ECG, EEG and EMG. However, the traditional analyses of these signals are currently not suitable with major advances in technology that allow for recording and storage of massive datasets of continuously fluctuating signals. Although these typically complex signals have recently been shown to represent processes that are non-linear, non-stationary, and non-equilibrium in nature, the methods used to analyse these data often assume linearity, stationary, and equilibrium-like conditions. In particular, studies have shown that sEMG signals may contain hidden information that is not extractable with conventional methods of analysis. Such hidden information will provide important and critical information to be of clinical value as well as to relate to basic mechanisms of muscle property and activity function.

References

1. Anmuth, C.J., G. Goldberg and N.H. Mayer. 1994. Fractal dimension of electromyographic signals recorded with surface electrodes during isometric contractions is linearly correlated with muscle activation. Muscle & Nerve, 17(8): 953–954.
2. Mandelbrot, B.B. 1977. Fractals: Form, chance, and dimension, first edn., W.H. Freeman and Co., San Francisco.
3. Gitter, J.A. and M.J. Czerniecki. 1995. Fractal analysis of the electromyographic interference pattern. Journal of Neuroscience Methods, pp. 103–108.
4. Gupta, V., S. Suryanarayanan and N.P. Reddy. 1997. Fractal analysis of surface EMG signals from the biceps. International Journal of Medical Informatics, pp. 185–192.
5. Hu, X., Z.Z. Wang and X.M. Ren. 2005. Classification of surface EMG signal with fractal dimension. Journal of Zhejiang University - Science, B 6(8): 844–848.
6. Arjunan, S.P. and D.K. Kumar. 2007. Fractal theory based non-linear analysis of sEMG. Proceedings of 3rd International Conference on Intelligent Sensors, Sensor Networks and Information Processing (ISSNIP), Melbourne, Australia, 545–548.
7. Arjunan, S.P. and D.K. Kumar. 2007. Fractal based modelling and analysis of electromyography to identify subtle actions. Proceedings of the 29th Annual International Conference of the IEEE EMBS, Lyon, France, 1961–1964.
8. Arjunan, S.P. and D.K. Kumar. 2010. Decoding subtle forearm flexions using fractal features of surface electromyogram from single and multiple sensors. Journal of NeuroEngineering and Rehabilitation, 7: 53.
9. Beretta-Piccoli, M., G. D'Antona, M. Barbero, B. Fisher, C.M. Dieli-Conwright, R. Clijsen and C. Cescon. 2015. Evaluation of Central and Peripheral Fatigue in the Quadriceps Using Fractal Dimension and Conduction Velocity in Young Females. PLoS ONE, 10(4): e0123921.
10. Arjunan, S.P. and D.K. Kumar. 2013. Age-associated changes in muscle activity during isometric contraction. Muscle Nerve, 47: 545–549.
11. Xu, Z. and S. Xiao. 1997. Fractal dimension of surface EMG and its determinants, in Engineering in Medicine and Biology Society, 1997. Proceedings of the 19th Annual International Conference of the IEEE, 4: 1570–1573.
12. Basmajian, J. and C.J. De Luca. 1985. Muscles Alive: Their Functions Revealed by Electromyography, fifth edn., Williams & Wilkins, Baltimore, MD.
13. Huang, H.-P. and C.-Y. Chen. 1999. Development of myoelectric discrimination system for a multi-degree prosthetic hand. IEEE International Conference on Robotic and Automation, 3: 2392–2397.
14. Dimitrov, G.V., T.I. Arabadzhiev, K.N. Mileva, J.L. Bowtell, N. Crichton and N.A. Dimitrova. 2006. Muscle fatigue during dynamic contractions assessed by new spectral indices. Med. Sci. Sports Exercise, 38: 1971–1976.
15. Dimitrova, N.A., T.I. Arabadzhiev, J.-Y. Hogrel and G.V. Dimitrov. 2009. Fatigue analysis of interference EMG signals obtained from biceps brachii during isometric voluntary contraction at various force levels. J. Electromyogr. Kines., 19(2): 252–258.
16. Kumar, D.K., S.P. Arjunan and G.R. Naik. 2011. Measuring Increase in Synchronization to Identify Muscle Endurance Limit. IEEE Transactions on Neural Systems and Rehabilitation Engineering, 19(5): 578–587.
17. Naik, G.R. and H.T. Nguyen. 2015. Nonnegative Matrix Factorization for the Identification of EMG Finger Movements: Evaluation Using Matrix Analysis. IEEE Journal of Biomedical and Health Informatics, 19(2): 478–485.
18. Chen, W.-T., Z.-Z. Wang and X.M. Ren. 2006. Characterization of surface EMG signals using improved approximate entropy. Journal of Zhejiang University – Science, B 7(10): 844–848.
19. Kleine, B.U., J.P. van Dijk, B.G. Lapatki, M.J. Zwarts and D.F. Stegeman. 2007. Using two-dimensional spatial information in decomposition of surface EMG signals. Journal of Electromyography and Kinesiology, 17(5): 535–548.

20. Lowery, M.M. and M.J. O'Malley. 2003. Analysis and simulation of changes in EMG amplitude during high-level fatiguing contractions. Biomedical Engineering, IEEE Transactions on, 50(9): 1052–1062.
21. Ancillao, A., M. Galli, C. Rigoldi and G. Albertini. 2014. Linear correlation between fractal dimension of surface EMG signal from Rectus Femoris and height of vertical jump. Chaos, Solitons & Fractals, 66: 120–126.
22. Kalden, R. and S. Ibrahim. 2004. Searching for self-similarity in GPRS, in PAM 2004: Passive and Active network Measurement, pp. 83–92.
23. Acharya, U., S.P. Bhat, N. Kannathal, A. Rao and C.M. Lim. 2005. Analysis of cardiac health using fractal dimension and wavelet transformation. ITBM-RBM, 26(2): 133–139.
24. Graupe, D. and W.K. Cline. 1975. Functional separation of SEMG signals via ARMA identification methods for prosthesis control purposes. IEEE Transaction on Systems, Man, and Cybernetics, 5(2): 252–259.
25. Durgam, V., G. Fernandes, H. Preiszl, W. Lutzenberger, F. Pulvermuller and N. Birbaumer. 1997. Fractal dimensions of short EEG time series in humans. Neuroscience Letters, 225(2): 77–80.
26. Peng, C.-K., J. Hausdorff and A. Goldberger. 1999. Fractal mechanisms in neural control: Human heartbeat and gait dynamics in health and disease. Nonlinear Dynamics, Self-Organization, and Biomedicine.
27. Bourke, P. 2007. Self similarity, Fractals, Chaos. URL: http://local.wasp.uwa.edu.au/pbourke/fractals/selfsimilar/.
28. Higuchi, T. 1988. Approach to an irregular time series on the basis of the fractal theory. Phys. D, 31(2): 277–283.
29. Katz, M.J. 1988. Fractals and the analysis of waveforms. Computers in Biology and Medicine, 18(3): 145–156.
30. Petrosian, A. 1995. Kolmogorov complexity of finite sequences and recognition of different preictal eeg patterns, in Proceedings of the Eighth IEEE Symposium on Computer-Based Medical Systems, pp. 212–217.
31. Esteller, R., G. Vachtsevanos, J. Echauz and B. Litt. 2001. A comparison of waveform fractal dimension algorithms. Circuits and Systems I: Fundamental Theory and Applications, IEEE Transactions on, 48(2): 177–183.
32. Boccia, G., D. Dardanello, M. Beretta-Piccoli, C. Corrado, G. Coratella, N. Rinaldo, M. Barbero, M. Lanza, F. Schena and A. Rainoldi. 2016. Muscle fiber conduction velocity and fractal dimension of EMG during fatiguing contraction of young and elderly active men. Physiol. Meas., 37(2016): 162–174.
33. Goldberger, A.L., L.A.N. Amaral, L. Glass, J.M. Hausdorff, P.C. Ivanov, R.G. Mark, J.E. Mietus, G.B. Moody, C.-K. Peng and H.E. Stanley. 2000. PhysioBank, PhysioToolkit, and PhysioNet: Components of a new research resource for complex physiologic signals. Circulation, 101(23): e215–e220.
34. Phinyomark, A., P. Phukpattaranont and C. Limsakul. 2012. Fractal analysis features for weak and single-channel upper-limb EMG signals. Expert Systems with Applications, 2012, 39(12): 11156–11163.

CHAPTER 6

Fractals Analysis of Electroencephalogram

ABSTRACT

Electroencephalogram (EEG) is the recording of the electrical activity from the scalp and changes in this signal have been associated with levels of alertness, and some disease conditions. It also finds applications for identifying motor commands directly from the brain.

Studies have demonstrated that EEG is largely a fractal signal, and changes to its fractal properties are associated with various brain conditions with applications such as identifying the mental alertness of the individual. The relationship of EEG fractal properties is also associated with other factors such as ageing. This chapter describes the fractal structure of the EEG signal in relations to the physiology and its application in the analysis of EEG. An example in the use of fractal dimension of EEG for measuring alertness is also provided.

6.1 Introduction

Electroencephalogram (EEG) is electrical activity recorded from the scalp. It is non-invasive recording and an indicator of the electrical activity in the brain which provides a measure of neuronal activity with high temporal resolution. EEG signal has been widely used as a valuable tool for the diagnosis and prognosis of many neurological disorders as well as for the monitoring of cerebral functions. The processing of the EEG signal is useful in:

a) Diagnosis of brain disorders by identifying abnormal patterns,
b) Identifying level of alertness,
c) Detection of neuropathic delays using event related potentials,
d) Determining brain commands.

6.1.1 History of EEG

The pioneering work in the recording of brain electrical activity was performed by English scientist Richard Canton using a galvanometer connected to the scalp of a human subject through two electrodes in 1875. The first reported continuous recording of the EEG was performed on photographic paper and was performed by a German psychiatrist Hans Berger in 1929 [1]. The generation mechanisms and the functional significance of the signals remained controversial for a long time as it was not able to relate the underlying cause of the signal. One hypothesis was that it was due to the complexity of the underlying systems of neuronal generators while the other was that it was because of the transfer of electrical charge from the cortical surface to the scalp due to the topological and electrical properties of the volume conductor [2].

6.1.2 Fundamentals of EEG

The EEG signal is a result of the potential induced on the scalp and because of the electrical activities of populations of neurons. These neurons are excitable cells with characteristic intrinsic electrical properties, and their activity produces electrical and magnetic fields which are recorded by means of electrodes located on the scalp.

The signal is a convenient way to study the brain. However, it is very gross and unspecific because it is the result of the summation of neural activity of large number of neurons. The other shortcoming is that the signal strength is small, typically around 1 microvolt, and the frequency is very low. This makes the signal affected by factors such as line noise, motion artefact and signal from other sources. To obtain useful information from the signal, it is essential to process the signal to reduce noise, and perform signal analysis to obtain suitable features.

Number of techniques have been developed to process and analyse EEG, some of these developed soon after discovering the signal. One of the earliest was performed on the early recordings, where Fourier transformation was applied to the recorded traces. As the techniques were developed over the years, the EEG signal as a source of information about the brain became more and more evident. This led to the development of clinical and experimental studies for detection, diagnosis, treatment and prognosis of several neurological abnormalities, as well as for the characterization of many physiological states.

The earlier analysis of EEG was largely based on the spectrum analysis. Perhaps the choice of the frequency bands was limited by the technology because of the relatively large size and low reliability of the electronic components. The signal was filtered in 5 spectral bands, in the range from

0.5 Hz to 60 Hz. This resulted in extensive studies that differentiated EEG in five bands.

The analysis in EEG signal led to the differentiation of the EEG signal into five major brain rhythms that have been recognized for over 6 decades. Each is characterized by a specific frequency range. The characteristics of the waves change not only with pathology or age, but also with the physiological state (for example alertness, sleep) [2,3].

- *Alpha*

  Alpha activity, in the range 8–13 Hz, is characteristic of eyes-closed awake state (relaxed awareness) and is mainly present in the occipital lobes.
- *Beta*

  The beta waves are associated with active thinking and problem solving and are usually found in adults within the range 13–30 Hz.
- *Gamma*

  Gamma activity refers to waves with frequencies above 30 Hz and seems to be related to consciousness.
- *Delta*

  The delta activity with the range 0.5–4 Hz, is primarily associated with deep sleep. Delta activity occurs also in case of coma and other disorders of consciousness, as well as during anesthesia.
- *Theta*

  Theta activity, with frequencies between 4 and 8 Hz, appears as consciousness slips toward drowsiness or during deep meditation. Theta waves occur mainly during infancy and childhood and are abnormal in the adults who are awake.

6.2 Techniques for EEG Analysis

There are a number of different methods for EEG analysis. This section discusses some of the major EEG analysis methods, and identifies the major measure of fractal properties of the signal.

In order to obtain meaningful information from the EEG signal, various techniques have been reported in decades of research. Spectral analyses are most important method in the extraction of information from the EEG signal to detect various states.

Spectrum estimation may be generally categorized into two classes [2,3].

(i) The nonparametric method (classical approach) deals with the estimation of the autocorrelation from a given data set. Fourier transformation is then applied to the estimated autocorrelation sequence to obtain the power spectrum. These methods suffer from spectral leakage effects due to windowing and masks weak signal components.

(ii) The parametric method (non-classical approach) is based on using a model for the process in order to estimate the power spectrum. Welch method is popularly used to estimate the power spectrum of a given time sequence. A parametric power spectrum estimation method overcomes the problem of spectral leakage and provides better frequency resolution since these methods assume the signal to be a stationary random process.

The spectral analysis of EEG has been found to be invaluable in discovering the functioning of the brain, for diagnosis of brain and nerve disorders and for developing brain command interface devices. However, these suffer from one major shortcoming; identifying the time when there is an event. Fourier transform is based on the assumption that the signal is stationary and its spectral properties do not change with time. While Fourier based analysis is suitable for identifying the stationary properties of the signal, this is unsuitable for monitoring changes over a period of time.

To monitor the change in the signal over time and to observe the time at which these spectral parameters change with respect to the mental states, time-frequency representations have been used. Short time Fourier Transform (STFT) Fourier Transform is one method that is suitable for time-frequency analysis. This method divides the signal into small temporal segments and the signal within this segment is assumed to be stationary. A window function is chosen whose width is equal to the segment of the signal, and the impulse response of which is suitable for the signal properties. While a rectangular window is easy to implement, this suffers from the ringing effect, it has been reported that STFT is not suitable for analyzing non-stationary signals because in STFT, a signal of finite length is expressed as the sum of frequency components of infinite duration. It fails to provide the exact location of an 'event' along the time scale in the frequency domain [4].

To overcome these limitations, wavelet transform techniques were introduced for time-frequency representations. Wavelet transform uses translations and dilations of a window function called 'wavelet'. There are two types of wavelet transforms (i) discrete wavelet transform (DWT) and (ii) continuous wavelet transform (CWT). These provide the signal spectral information without sacrificing the temporal data by performing convolution with the wavelet functions. Wavelet is the term used to define the 'baby wave' which are waves that are localised in time and frequency domain. These can also be considered as a combination of time windowed spectral filters and provide the flexibility of the shape function and length of the coefficient. However, that flexibility comes at the cost of matching the wavelet function to the signal characteristics.

Other ways of obtaining the spectral information of EEG is based on the computation using higher order statistics (HOS), i.e., moments and cumulants of third and higher order. The Bispectrum quantifies the relationship among the underlying components of the EEG and is very useful for analysing signals such as EEG which are not truly Gaussian. This is suitable for detecting the quadratic phase coupling between distinct frequency components in EEG signals [5].

The other important techniques used for the analysis of EEG are based on signal decomposition and prediction methods such as independent component analysis (ICA) which is being used to perform blind source separation of the signal originating from different sources. While the components within EEG may not be strictly independent, ICA has been found to be very useful to separate independent sources linearly mixed in several sensors. For example, ICA can separate the artefacts embedded in the EEG data as they are usually independent of each other. Research studies have used ICA to remove artefacts from EEG for extraction of meaningful features. All these techniques have been used to extract information related to the functional states. However, it is difficult to relate these to the underlying physiological phenomenon.

One important property of the signal is its complexity and chaotic behaviour. Having similar sources, the neurons, makes the inherent self-similarity an obvious property of the signal, making it suitable for being measured using fractal properties. Many studies have identified the fractal properties as an important measure of the neurological processes underlying EEG.

6.3 Fractal Properties of EEG

Electroencephalogram is influenced by the level of alertness of the person, and hence its analysis provides a method to determine the modification of the brain activities at different stages of alertness and during the transition from wakefulness to sleep or with age. This has been found to be a very useful application of EEG and many studies have attempted to identify suitable signal processing techniques and signal features to analyse and classify EEG recordings. It has been commonly considered that EEG is based on stationary stochastic processes and analysed by linear techniques. Other methods used for the analysis are; autocorrelation of EEG samples, and the power within spectral bands and these have been widely used as quantitative measures of brain electrical activity [1,2].

More recent viewpoint to address the irregular behaviour of the EEG arises from chaos theory. Based on the chaos theory, randomness can also be displayed by nonlinear dynamical systems with limited degrees of freedom. Chaotic systems, though deterministic, are highly unpredictable due to their sensitive dependence on initial conditions. However, the rules governing the dynamics of such systems are, actually, simple. Chaos theory provides new techniques for the analysis of many physiological systems and suggests the existence of simple mathematical models for their description [6,7].

The most characteristic measures of a chaotic system are

a) the largest Lyapunov exponent, which quantifies the rate of the exponential divergence of nearby trajectories in the phase space, and
b) the correlation dimension, representing the unusual geometry [6]

The most reliable way to approach the study of complex systems is by fractal geometry. As explained in the previous chapters, the main features of fractal objects, namely self-similarity and non-integer dimension which can be displayed directly in the time domain. Based on these characteristics, the fractal-like behavior of the EEG and its unusual power spectrum can be characterized by parameters like the fractal dimension, the power-law exponent and the Hurst index [1,5].

Researchers have reported that EEG waveforms corresponding to different physio-pathological conditions can be characterized by their complexity and randomness [3,7,8,10]. One measure of complexity of a signal is the fractal dimension (FD), and the advantage of this is that it is a global property of the signal, making it very suitable for machine based analysis. The use of the FD to analyze EEG signal has been reported since 1990s. Studies by [7,9] have demonstrated the fractal nature of EEG and its association with factors such as sleepiness, alertness and diseases.

Klomwski et al. [11] applied chaos theory and fractal dimension of EEG signal recorded from patients with seasonal affective disorder. Arjunan et al. [9] have used fractal features of EEG to measure the alertness level of the user while driving. Raghavendra et al. [11] reported that schizophrenic patients had a lower FD of bifrontal EEG when compared with healthy control.

Uthayakumar et al. [6] have used fractal dimension of EEG signal to differentiate between healthy and epileptic signal and have also introduced a new modified method to compute fractal dimension. In the next section, we will discuss an example of analysis of EEG [9] using fractal theory.

6.4 An Example—Measuring Alertness Using Fractal Properties of EEG

Alertness deficit is a major problem where the operator is monitoring powered equipment such as automobile, farm machinery or factory equipment or for personnel responsible for controlling complex situations such as defence personnel. People who are undertaking tasks which are often monotonous such as long-distance drivers or sentry in remote locations or people performing a repetitive task over extended periods of time tend to become distracted from the task and lose their alertness. It has been the cause of major catastrophic events for people driving vehicles, monitoring power plant, participating in defence maneuvers and even people monitoring security or computer screens for extended periods of time. Earlier studies have shown that retaining a constant level of alertness is difficult or impossible for operators of motorized systems.

There is an urgent need for non-invasively detecting the loss of alertness and two modalities that have been tested successfully are; eye gaze and EEG. While eye-gaze has the advantage that it can be monitored using a camera, it suffers from the need for appropriate lighting conditions and for concerns of reliability due to the movement of the person. EEG has the shortcoming that there is the need for fixing electrodes on the scalp of the person, and traditionally EEG devices have been bulky and with large number of wires. However, with the significant improvement in electronics and wireless capabilities, new EEG devices have smart headsets with wireless connectivity that can be mounted by the users themselves.

Research studies have shown the relationship of Electroencephalogram (EEG) with changes in alertness, arousal, sleep and cognition [13–18] Study by Jung et al. [19,20] has estimated alertness of people using power spectrum of EEG. However, one shortcoming with biosignals such as EEG is the very low signal to noise ratio. The typical signal strength of EEG signal is of the order of 1 micro-volt, and often the strength of artifacts and noise may be much greater than this. Artifacts such as electro-ocular gram (EOG) can often be an order of magnitude that is greater, making the use of EEG for automated analysis difficult and unreliable.

The other major shortcoming is when wireless headset based EEG devices are used and self-mounted by the user, there can be large variations in the amplitude of the signal. For reliable assessment of the signal, it is essential that the features be insensitive to the amplitude, and be noise tolerant. Fractal dimension of EEG is a global feature of the signal that is a measure of its complexity and not highly dependent on the signal amplitude. This section demonstrates the effectiveness in the use of FD of EEG to identify when the driver has reduced alertness.

6.4.1 Experimental setup

The experiment was conducted at Swartz center and has been reported in publications [19,20]. During the experiment, EEG was recorded while the volunteer drove a simulated car. During the simulation, obstacles were introduced and the response time of the volunteer to avoid the obstacle was measured and inversely associated with the alertness. This experimental data was obtained from Swartz Center for Computational Neuroscience, Institute for Neural Computation, University of California, San Diego. The experimental setup explained in this section has been obtained with permission from their referred publications [19,20] and has been acknowledged.

6.4.2 Subjects

Three healthy subjects (ages from 18 to 34) participated in a dual-task simulation of auditory sonar target detection. All had passed the standard US Navy hearing tests and reported having normal hearing. Each subject participated in three or more simulated work sessions each lasting 28 minutes. Each participant was given an oral and written summary of the experimental protocol.

6.4.3 Stimuli

Auditory signals, including background noise, tone pips, and noise burst targets, were synthesized using a concurrent work station which was also used to record the EEG. In a continuous 63-db white noise background, task-irrelevant auditory tones at two frequencies (568 Hz and 1098 Hz) were presented in random order at 72dB (normal hearing level) with stimulus onset asynchronies between 2–4s.

These signals were introduced to assess the information available in event-related potentials [21] and are not reported in this study. In half of the inter-tone intervals, target noise bursts were presented at 6dB above their detection threshold and the mean target rate was thus 10 per minute.

6.4.4 EEG recording and processing

EEG data were recorded at a sampling rate of 312.5 Hz from two midlines sites, one central (Cz) and other midway between parietal and occipital sites (Pz/Oz), using 10-mm gold-plated electrodes located at sites of the International 10–20 system, referenced to the right mastoid. EEG data were first preprocessed using a simple out-of-bounds test (with a 50μV threshold) to reject epochs that were grossly contaminated by muscle and/or eye-movement artefacts'. Moving averaged spectral analysis of the EEG data was then accomplished using a 256-point Hanning-window with 50% overlap. Windowed 256-point epochs were extended to 512 points by zero-padding. Median filtering using a moving 5-s window was used to further minimize the presence of artifacts in the EEG records. Two sessions from each from the three of the participants were chosen for analysis on the basis of their including more than 50 detection lapses.

6.4.5 Experimental procedure

Experimental procedure was designed in order to determine the level of alertness from EEG recordings. Each subject participated in three or more 28-min experimental sessions on separate days. During the experiment, the participants mimicked audio sonar target detection. The participants were asked to respond to given auditory commands. The subjects pushed one button whenever they detected an above-threshold auditory target stimulus (a brief increase in the level of the continuously-present background noise). To maximize the chance of observing alertness decrements, sessions were conducted in a small, warm and dimly-lit experimental chamber, and subjects were instructed to keep their eyes closed.

6.4.6 Alertness measure

Auditory targets were classified as Hits or Lapses depending on whether or not the subject pressed the auditory response button within 100 ms to 3000 ms of target onset. To quantify the level of alertness, auditory responses were converted into local error rate, defined as fraction of targets not detected by the subject (i.e., lapses) within a moving time window. A continuous measure, local error rate, was computed by convolving an irregularly spaced performance index (hit = 0/lapse = 1) with a 95s smoothing window advanced through the performance data in 1.64s steps. Each error rate time series consisted of 1024 points at 1.64s intervals. Error rate and EEG data from the first 95s of each run were not used in the analysis. For each window position, the sum of window values at moments of presentation of undetected (lapse) targets was divided by the sum of window values at moments of presentation of all targets. The window was moved through the session in 1.64s steps, converting the irregularly-sampled, discontinuous performance record into a regularly-sampled, continuous error rate measured with range [0,1].

6.4.7 Data analysis

In order to perform fractal analysis, the fractal features were computed as explained in Chapter 3, Section 3.3. The fractal features Maximum Fractal Length (MFL) and Fractal Dimension (FD) were computed from the EEG data using a stepping window of 1.64s and were analyzed to determine the correlation with the local error rate. The results of the experiments were analyzed to determine the alertness levels in relation to the small changes in EEG using the correlation analysis.

Correlation analysis

Correlation analysis often measured as a correlation coefficient, indicates the strength and direction of a linear relationship between two random variables. In general statistical usage, correlation or co-relation refers to the departure of two variables from independence. In this broad sense there are several coefficients, measuring the degree of correlation, adapted to the nature of data. In this context, the null hypothesis asserts that the two variables are not correlated, and the alternative hypothesis asserts that the attributes are correlated.

Correlation coefficient

The correlation coefficient r is a measure of the linear relationship between two attributes or columns of data. The correlation coefficient is also known as the Pearson product-moment correlation coefficient. The value of r can range from –1 to +1 and is independent of the units of measurement. A value of r near 0 indicates little correlation between attributes; a value near +1 or –1 indicates a high level of correlation.

When two attributes have a positive correlation coefficient, an increase in the value of one attribute indicates a likely increase in the value of the second attribute. A correlation coefficient of less than 0 indicates a negative correlation. That is, when one attribute shows an increase in value, the other attribute tends to show a decrease.

Consider two variables x and y:

- If $r = 1$, then x and y are perfectly positively correlated. The possible values of x and y all lie on a straight line with a positive slope in the (x,y) plane.
- If $r = 0$, then x and y are not correlated. They do not have an apparent linear relationship. However, this does not mean that x and y are statistically independent.
- If $r = -1$, then x and y are perfectly negatively correlated. The possible values of x and y all lie on a straight line with a negative slope in the (x,y) plane.

6.4.8 Discussion

In this EEG analysis, the MFL and FD data points were determined using stepped window of 1.6s, and were fitted using polynomial fit for each session. The error rate corresponding to each session was fitted using same polynomial function. To determine the relation between the changes in MFL with error function, the correlation coefficients were calculated. The correlation coefficients as a measure show the performance of MFL & FD of EEG in relation with the level of alertness.

Figure 6.1 shows the plot of correlation between the fractal dimension, MFL and error rate function. From the plot, it is observed that using fractal features it is reliable to determine or indicate the level of alertness. The fractal features were correlated (negative) with corresponding local error rate to determine the alertness measure. The negative correlation coefficients between the MFL & FD with error function are shown in Table 6.1. The results demonstrate that the fractal features and local error rate have high negative correlation coefficients (Mean = 0.8196/SD = 0.02810).

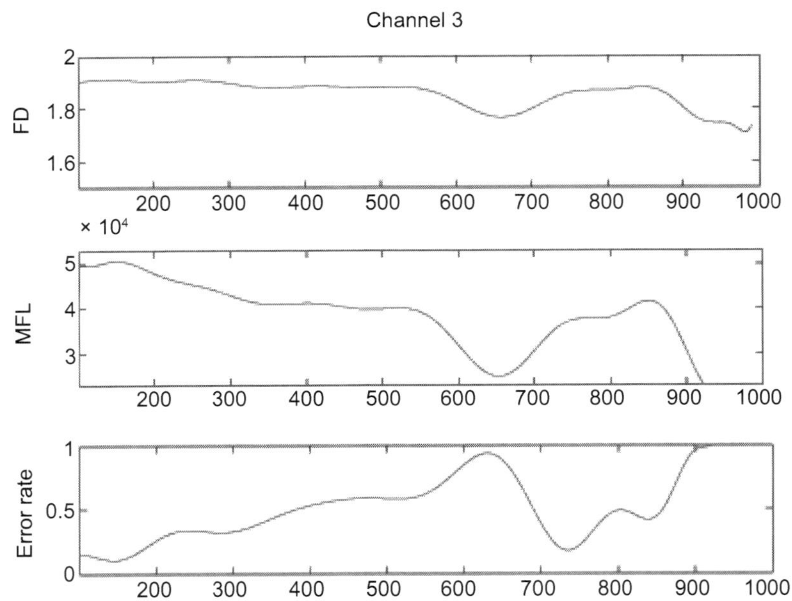

Figure 6.1. Plot of FD and MFL (from two channels during a session) inversely correlated with the local error rate using polynomial fit.

The results indicate that MFL is linearly and inversely correlated with the fluctuations of the subject's task performance and putative alertness level with mean negative correlation coefficient of 0.82. It is also observed that other features such as fractal dimension and PSD also correlate with the subject's task performance, but to a lesser degree. The correlation accuracy of FD was 0.73 while that of PSD was 0.77.

Table 6.2 shows the correlation coefficients of the predicted and actual local error rate. The predicted error rate was calculated based on the changes in the MFL of EEG. The results also indicate that the changes in the alertness level can be predicted using MFL of EEG.

Table 6.1. Correlation coefficients for MFL, FD and PSD with the local detection lapses.

Exp. Nos.	MFL (one channel)	FD (one channel)	PSD (8 channels)
Subject A 1	–0.84	–0.76	+0.87
2	–0.80	–0.73	+0.73
Subject B 3	–0.83	–0.74	+0.83
4	–0.81	–0.72	+0.76
Subject C 5	–0.80	–0.71	+0.76
6	–0.79	–0.69	+0.70

Table 6.2. Correlation coefficients of MFL of single and two channel for session 1 and 2 (same session and different sessions). Session 2 correlation is a measure of prediction of detection of lapses.

	Testing	Prediction	Testing	Prediction
	Single Channel		Two Channels	
Session	Subject 1			
1	0.84	0.65	0.91	0.71
2	0.81	0.64	0.89	0.68
Session	Subject 2			
3	0.83	0.60	0.88	0.69
4	0.81	0.54	0.89	0.62
Session	Subject 3			
5	0.80	0.62	0.89	0.70
6	0.79	0.63	0.87	0.71
Average	0.81	0.61	0.89	0.69

The data was also analyzed for the inter session variations for the different participants as shown in the bar plot (Fig. 6.2) and in Tables 6.1 and 6.2. This bar plot shows that the negative correlation coefficients are more similar in two sessions. It suggests that the MFL is reliably correlated with error function in two sessions. The alertness changes are reliably measured during the two sessions using the minimum number of channels (in this case two channels).

This study has identified changes in the fractal features of EEG recordings in response to the changes in alertness of the subject. It has demonstrated that in comparison with PSD and FD, MFL correlates best

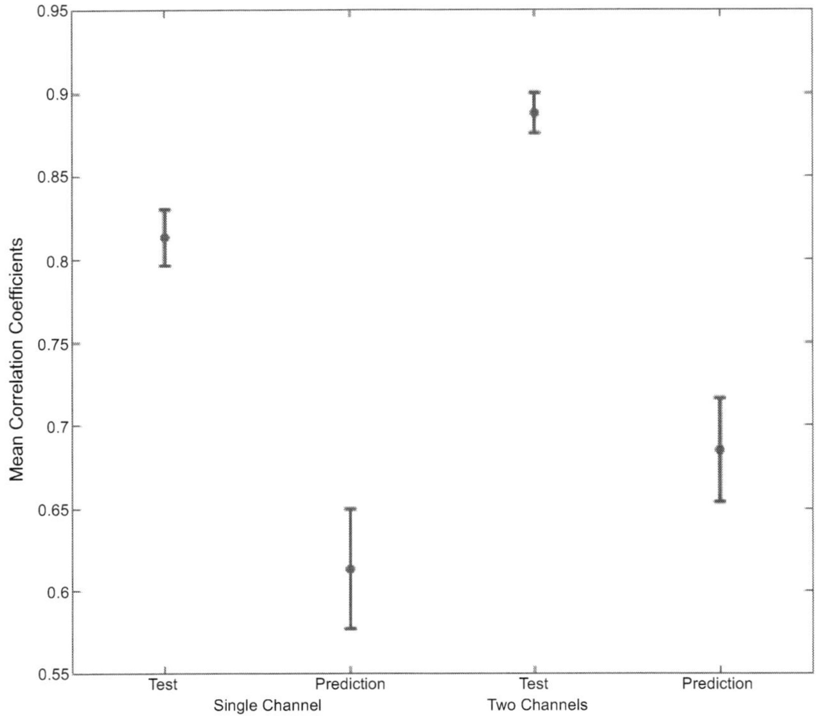

Figure 6.2. Mean correlation coefficients of the prediction of the alertness level using the MFL of EEG from tested session.

with the alertness of the subject. It is important to understand that the properties of the EEG signal will change based on the mental states. In this example, the fractal dimension reduces as there is a change in alertness state which in turn changes the randomness associated with the recorded EEG signal. This performance analysis demonstrates that it is feasible to use fractal features of only two channels of EEG to track an operator's global level of alertness in a sustained-attention task.

6.5 Summary

This chapter has discussed the fractal properties of EEG in order to extract the randomness in the structure which provides an additional approach to analyze the signal. The unusual behavior of irregular time series such as EEG signal can be analyzed using the fractal theory which is based on the irregular pattern and complexity. An example to observe the fractal nature of EEG to correlate with the mental state of alertness has been discussed.

References

1. Cusenza, M. 2011. Fractal Analysis of the EEG and Clinical Applications, PhD Thesis, Retrieved from https://www.openstarts.units.it/dspace/bitstream/10077/7394/1/cusenza_phd.pdf.
2. Lopes da Silva, F. 2010. EEG: Origin and Measurement EEG–fMRI. C. Mulert and L. Lemieux (eds.), 19–38.
3. Durgam, V., G. Fernandes, H. Preiszl, W. Lutzenberger, F. Pulvermuller and N. Birbaumer. 1997. Fractal dimensions of short eeg time series in humans. Neuroscience Letters, 225(2): 77–80.
4. Leonowicz, Z., T. Lobos and K. Wozniak. 2009. Analysis of non-stationary electric signals using the S-transform. COMPEL—The International Journal for Computation and Mathematics in Electrical and Electronic Engineering, 28(1): 204–210.
5. Goshvarpour, A., A. Goshvarpour, S. Rahati and V. Saadatian. 2012. Bispectrum estimation of electroencephalogram signals during meditation. Iranian Journal of Psychiatry and Behavioral Sciences, 6(2): 48–54.
6. Paramanathan, P. and R. Uthayakumar. 2008. Application of fractal theory in analysis of human electroencephalographic signals. Computers in Biology and Medicine, 38(3): 372–378.
7. Accardo, A., M. Affinito, M. Carrozzi and F. Bouquet. 1997. Use of the fractal dimension for the analysis of electroencephalographic time series. Biological Cybernetics, 77(5): 339–350.
8. Beckers, F., B. Verheyden, K. Couckuyt and A.E. Aubert. 2006. Fractal dimension in health and heart failure. Biomedizinische Technik. Biomedical Engineering, 51(4): 194–197.
9. Arjunan, S.P., D.K. Kumar and T.P. Jung. 2010. Estimation of alertness levels with changes in decibel scale wavelength of EEG during dual-task simulation of auditory sonar target detection. Annual International Conference of the IEEE Engineering in Medicine and Biology Society (EMBC), 2010, 4444–4447.
10. Subha, D.P., P.K. Joseph, R.U. Acharya and C.M. Lim. 2010. EEG signal analysis: A survey. J. Med. Syst., 34: 195–212.
11. Klonowski, W., W. Jernajczyk, K. Niedzielska, A. Rydz and R. Stepień. 1999. Quantitative measures of complexity of EEG signal dynamics. Acta Neurobiol. Exp. (Wars), 59(4): 315–21.
12. Raghavendra, B.S., D.N. Dutt, H.N. Halahalli and J.P. John. 2009. Complexity analysis of EEG in patients with schizophrenia using fractal dimension. Physiol. Meas, 30: 795–808.
13. Bullock, T.H., M.C. McClune, J.Z. Achimowicz, V.J. Iragui-Madoz, R.B. Duckrow and S.S. Spencer. 1995. Temporal fluctuations in coherence of brain waves. Proceedings of the National Academy of Sciences of the United States of America, 92(25): 11568–11572.
14. Chapotot, F., C. Gronfier, C. Jouny, A. Muzet and G. Brandenberger. 1998. Cortisol secretion is related to electroencephalographic alertness in human subjects during daytime wakefulness. The Journal of Clinical Endocrinology and Metabolism, 83(12): 4263–4268.
15. Grigg-Damberger, M., D. Gozal, C.L. Marcus, S.F. Quan, C.L. Rosen, R.D. Chervin, M. Wise, D.L. Picchietti, S.H. Sheldon and C. Iber. 2007. The visual scoring of sleep and arousal in infants and children. Journal of Clinical Sleep Medicine: JCSM: Official Publication of the American Academy of Sleep Medicine, 3(2): 201–240.
16. Makeig, S. and M. Inlow. 1993. Lapses in alertness: coherence of fluctuations in performance and EEG spectrum. Electroencephalography and Clinical Neurophysiology, 86(1): 23–35.
17. Silber, M.H., S. Ancoli-Israel, M.H. Bonnet, S. Chokroverty, M.M. Grigg-Damberger, M. Hirshkowitz, S. Kapen, S.A. Keenan, M.H. Kryger, T. Penzel, M.R. Pressman and C. Iber. 2007. The visual scoring of sleep in adults. J. Clin. Sleep Med., 3(2): 121–131.
18. Tassi, P., A. Bonnefond, O. Engasser, A. Hoeft, R. Eschenlauer and Muzet. 2006. EEG spectral power and cognitive performance during sleep inertia: the effect of normal sleep duration and partial sleep deprivation. Physiology & Behavior, 87(1): 177–184.

19. Jung, T.-P., S. Makeig, M. Stensmo and T.J. Sejnowski. 1997. Estimating alertness from the eeg power spectrum. Biomedical Engineering, IEEE Transactions on, 44(1): 60–69.
20. Makeig, S., T.-P. Jung and T.J. Sejnowski. 1996. Using feed-forward neural networks to monitor alertness from changes in eeg correlation and coherence. Advances in Neural Information Processing Systems, 8: 931–937.
21. Venturini, R., W. Lytton and T. Sejnowski. 1992. Neural network analysis of event related potentials and electroencephalogram. Advances in Neural Information Processing Systems, 4: 651–658.

CHAPTER 7

Fractal Analysis of Biomedical Images

ABSTRACT

The fractal and entropy concepts have been widely used for analysis of medical images. This is due to the unique characteristics of such images that show a certain degree of randomness associated with properties of anatomical structures. There are a number of algorithms that have been developed for measuring the fractal dimensions (FD). In order to better understand the terms and practical application of each method, this chapter first introduces the state of the art techniques which have been mostly used for biomedical image analyses. It then evaluates the literature to identify the differences between various approaches and provides the readers with detailed explanations for easy implementation of each method in practice.

7.1 Introduction

Medical image analysis encompasses various approaches for extraction and processing of diagnostically valuable information, using medical images that are taken with different imaging modalities. The goal is to quantitatively assess a hypothetical link between a disease abnormality (pathology), progression and the change in its properties over time such as the size, complexity and texture which may or may not be visually observable. Any detectable change within a tissue and its surrounding area could be a key component in every diagnostic problem. One concept that has been effectively used in medical image analysis and study of association between a disease incidence and its risk factors is the fractal dimension (FD). FD is often interpreted as global measure of the complexity and irregularity of a geometry which is useful for quantifying and summarizing anatomical structures and monitoring the changes over time. For instance overall reduction in FD has been found to be consistent with reduction in complexity of retinal vasculature as a result of ageing [1,2]. Therefore reduced fractal dimension of retinal vasculature can imply reduction in or loss of branching pattern as a result of healthy aging. As another example, studies have shown that there is a link between changes in retinal vasculature morphology and risk of cerebral infarction and have used FD to non-invasively quantify the changes in the complexity of retinal vessel and to assess the risk of a stroke event. For instance, in a population based cohort study by Kawasaki et al., each Standard Deviation (SD) decrease in FD value of retinal vessels was found to be inversely associated with 40% higher risk of stroke event [3].

This chapter will explore such concepts and methods used for quantifying the complexity, irregularity and randomness of natural geometries such as anatomical structures in medical images. Throughout this section, retinal vasculature has been used as example of a biomedical structure that exhibits fractal properties in order to introduce the mathematical concepts. Although retinal vasculature is known as semi-fractal geometry, FD has been widely accepted and employed by many studies for retinal image analysis. It should be noted that all the techniques and mathematical concepts presented in this chapter are applicable to other medical images with fractal or semi-fractal properties.

7.2 Fractal Geometry and Self-similarity

Fractal geometry is a concept which mathematically originates from the problem of quantifying the dimension of irregular geometries; and is characterized by self-similarity properties. A self-similar object is referred to a geometrical shape that consists of exact or approximate copies of itself in each of its spatial directions when reduced into smaller scales. Examples of self-similar fractal geometries include shape of a coastline, Sierpinski Triangle, Koch snowflake and fern leaf. From the mathematical point of view, if a self-similar object is successively reduced by a factor of $\frac{1}{x}$ in each of its spatial directions, it will consist of x^D reduced-scale replications of itself with D being the dimension of the object. For Euclidean and regular geometries such as lines, squares and cubes D will take integer values of one, two or three respectively. However, shape of a fractal geometry as an irregular object, cannot be simply described by Euclidean dimension. For instance, topological dimension of a coast line cannot be addressed by the concept of Euclidean geometry. Such geometry will have unbounded lengths extending towards infinity when looking deep into its microscopic details. Also the length gets longer and longer as we go deeper into its details. In order to measure the length of the coastline, one should measure the length of every rock, pebble and grain of sand. However, such measurement will still be an approximation of the actual length. Therefore, there is a need for definition of a different concept to better quantify the progression of details in fractal geometry and to describe the shape. This is why FD comes into play as a useful measure to summarize the shape of fractal geometries. It can take non-integer values meaning that unlike the Euclidian objects, fractal geometries such as the coast line is neither 1D nor 2D but its dimension is a value between 1 and 2.

7.3 Entropy, Fractals and Tortuosity

The notion of entropy was first introduced in classical thermodynamics as quantitative measure of the evolution of a system with time. The term was used as the basis for the second law of thermodynamic which states that the entropy of an isolated system will increase as long as the energy of the system is transformed from one type to another and remains constant when the system is in equilibrium state. Therefore it either always increases or stays the same and unlike energy, which cannot be created and is only transformed from one type to another, entropy can be created. It is often referred to as the measure of disorder, randomness, or multiplicity of a system; but, still there is no clear definition for entropy and the concept has been employed intuitively in many other areas such as image processing. Now, the question is how entropy can be interpreted in image processing and how the degree of disorder or randomness can be measured and applied to real world analyses?

In image processing, entropy provides statistical information regarding the uniformity of distribution of the gray level intensities in an image. It is a feature, based on either first or second order statistics from gray level histogram that characterizes textural properties for image classification and pattern recognition. The mathematical definition of entropy using the first order histogram is expressed as

$$entropy = -\sum_{i=0}^{G-1} p(i)\log(p(i)) \tag{7.1}$$

where $p(i)$ is the ith element of the first order statistics from gray level histogram of the image with G being the number of possible gray levels. $p(i)$ is defined as the ratio of the total number of pixels with gray level i to the total number of pixels in the image.

Entropy obtained from the first-order statistics provides no information about the spatial relationship between the pixels within the gray-scale image. However, the spatial distance between the pixels and their relative orientation are of importance. Therefore instead of using first order statistic histogram, entropy is often derived from the second order statistics using gray-level spatial-dependent matrix, also known as gray-level co-occurrence matrix (GLCM). It shows the frequency of the intensity value i within an image that occurs in a specific spatial relationship (distance and orientation) to a pixel with the value *j*. Assuming a window size of $w \times w$ with G quantized gray levels, a $G \times G$ GLCM matrix can be obtained by choosing a distance (d) and orientation (θ) and counting the number of pixel pairs where the first and second pixels in the pair have values of *i* and *j* respectively. The values are then normalized by the number of pixel pairs in the image to get the co-occurrence probabilities. θ is quantized in four different angles of 0°, 45°, 90° and 135° but the choice of d, G and w is usually made heuristically (typically $d = 1$ to 3, $G = 16$, $w = 30$ to 50). Figure 7.1 shows selection of distance and orientation from a candidate pixel for obtaining GLCM matrix.

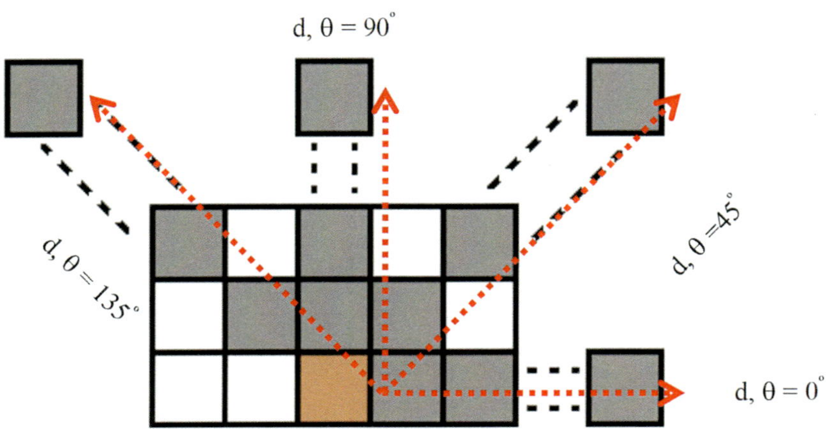

Figure 7.1. Illustration of spatial distance (d) and orientation (θ) from the candidate pixel for calculation of GLCM matrix.

As an example, assuming I is a window of size $w \times w$ obtained from a section of an image. Considering $d = 1$, $\theta = 0°$ and 4 levels of grayscale quantization, the GLCM matrix and co-occurrence probabilities can be obtained as follow:

$$I = \begin{bmatrix} 33 & 32 & 33 & 34 & 32 & 33 & 34 \\ 36 & 38 & 36 & 38 & 38 & 39 & 40 \\ 59 & 61 & 59 & 60 & 60 & 58 & 60 \\ 82 & 86 & 79 & 79 & 80 & 76 & 79 \\ 76 & 78 & 75 & 74 & 74 & 78 & 80 \\ 54 & 55 & 55 & 51 & 52 & 53 & 53 \\ 39 & 40 & 40 & 35 & 38 & 36 & 38 \end{bmatrix}_{w \times w} \xrightarrow[(G=4)]{\textit{Quantization of Gray Levels}} \begin{bmatrix} 0 & 0 & 0 & 0 & 0 & 0 & 0 \\ 0 & 0 & 0 & 0 & 0 & 0 & 0 \\ 2 & 2 & 2 & 2 & 2 & 1 & 2 \\ 3 & 3 & 3 & 3 & 3 & 3 & 3 \\ 3 & 3 & 3 & 3 & 3 & 3 & 3 \\ 1 & 1 & 1 & 1 & 1 & 1 & 1 \\ 0 & 0 & 0 & 0 & 0 & 0 & 0 \end{bmatrix}_{w \times w} \xrightarrow[(d=1,\,\theta=0°)]{\textit{GLCM Matrix}}$$

$$\begin{bmatrix} 18 & 0 & 0 & 0 \\ 0 & 6 & 1 & 0 \\ 0 & 1 & 4 & 0 \\ 0 & 0 & 0 & 12 \end{bmatrix}_{G \times G} \xrightarrow{\substack{\textit{normalized GLCM Matrix} \\ \textit{(co–occurrence probabilities)}}} p = \frac{1}{42} \begin{bmatrix} 18 & 0 & 0 & 0 \\ 0 & 6 & 1 & 0 \\ 0 & 1 & 4 & 0 \\ 0 & 0 & 0 & 12 \end{bmatrix}$$

According to this example there are a total number of 18 pixels in the quantized input image where two horizontally adjacent pixels (i.e., $\theta = 0°$) with distance $d = 1$, have the values $i = 0$ and $j = 0$ respectively. Therefore 18 is placed in the $(i = 0, j = 0)^{th}$ element of the $G \times G$ GLCM matrix where $G = 4$ equivalent to the number of quantization levels. Similarly, the $i = 2^{th}$ and $j = 2^{th}$ element of the GLCM matrix shows there are 6 candidate pixels in total with intensity value 2 located next to another pixel with quantized gray level of 2. The co-occurrence probability is obtained by dividing each element by the sum of all the elements in the GLCM matrix.

The mathematical definition of entropy using second order statistics and co-occurrence probabilities is as follow:

$$entropy = -\sum_{i=0}^{G-1}\sum_{j=0}^{G-1} p(i,j)\log(p(i,j)) \tag{7.2}$$

where $p(i,j)$ is the $(i,j)^{th}$ element in the GLCM matrix with G being the total number of possible gray levels. In other words, $p(i,j)$ is the 2nd order joint probability of changing from gray level i to j with respect to a fixed distance and orientation $(p(i,j))$. The less-smooth the image is, the more uniformly $p(i,j)$ will be distributed.

In addition to the entropy, which describes the degree of disorder or randomness, attempts have been made to formulate the degree of tortuosity and curvature of a tubular structure. This is a useful measure to quantify how much a curve is twisted compared to a straight light. Such measurement has found many real world applications from estimation of diffusion in a porous media such as soil or snow to medical image analyses for study of pathological changes or effect of treatments. When it comes to retinal image processing, vessel tortuosity is used as indicator of arterial hypertension, diabetic retinopathy, cerebral vessel disease, stroke and Retinopathy of Prematurity (ROP) a condition of vascular abnormalities found in the premature infants. Vessel tortuosity in a simple way is defined as the actual path length of the vessel segment divided by the Euclidian distance between the two end points of the segment. Considering skeletonized retinal images, simple tortuosity τ can be calculated as the ratio of the length of the vessel segment L to the Euclidean distance between two end points of a vessel segment (chord length) C as in eqn. 7.3. The minimum value that τ can take is equal to one (i.e., $\tau \geq 1$) which is for the cases of straight vessels when $L = C$.

$$\tau = \frac{L}{C} \tag{7.3}$$

It should be noted that this is a very basic definition of tortuosity which has a number of drawbacks. For instance it is quite possible to find two vessels with the same L and C that differ in the number of curves. In this case, the eqn. 7.3 will give similar values for both cases. However, in reality the two vessel segments can have different curvature properties.

To overcome this problem a number of methods have been developed for tortuosity measurement which are beyond the scope of this book.

As another measure, FD was introduced by Benoit Mandelbrot [4] by drawing attention to the length of the coastline of Great Britain. He argued that the length of the coastline can have different interpretations depending on the estimation method. Mandelbrot suggested using FD as measure of the complexity of a geometry such as the coastline by evaluating how fast the length would increase with respect to reduction in scale [5]. There are many established techniques for FD calculation of biomedical images and have been extended to 2D and 3D applications; including Binary Box-counting, Differential 3D Box-counting [6], Fourier (Spectral) Fractal [7] and Higuchi's method [8]. The rest of this chapter will provide details of the above techniques which have been mostly used in medical image analyses.

In summary the entropy, fractal dimension and even tortuosity are useful parameters to characterize biomedical images such as retinal vasculature and monitor the morphological changes to natural structures which can happen as a result of disease incidence.

7.4 Binary Box-count Fractal Dimension

There are many established versions of Box-counting algorithms available for fractal analysis of fractal geometries [9]. Binary Box-counting FD which is a special form of Mandelbrot's fractal dimension, has been frequently used for analysis of any structure in the plane or space. Also it is generally accepted for retinal image analyses by earlier works for quantifying how the detail in retinal vasculature pattern changes with change in scale (known as complexity) [10–12]. This method is only applicable to binary images and requires image segmentation as a pre-processing step. It involves superimposing the structure (vascular network in retinal images) with a grid of boxes with varying (either decreasing or increasing) side length of size R and counting the number of boxes N corresponding to each R; which contain at least one pixel of the structure. The slope of the best fitting line to the data points on a *Log-Log* plot of $Log(N)$ vs. $Log(\frac{1}{R})$ corresponds to Box-Counting fractal dimension as in eqn. 2.6, which is a number between 1 and 2 for a plane.

$$FD_{BC} = \frac{Log(N)}{Log(\frac{1}{R})} \tag{7.4}$$

An example of mesh grid superimposition on a binarized retinal image and the estimated Box-Counting dimension is shown in Fig. 7.2. The boxes that contain pixels from the vessel structure are shown in red. In this image, the pixels are partitioned into two groups of black and white pixels corresponding to the background and vascularized area respectively while preserving the vessel diameter information. However, for analysis of images containing vascular or any tubular structure such as retina, the vessel width information is often removed prior to FD calculation by using a technique called "skeletonization". It is a morphological operation applied directly to the binary image which provides approximation of the vessel centre lines. It refers to the process of obtaining a single pixel version of the vascular network corresponding to a line that is equidistant from vessel.

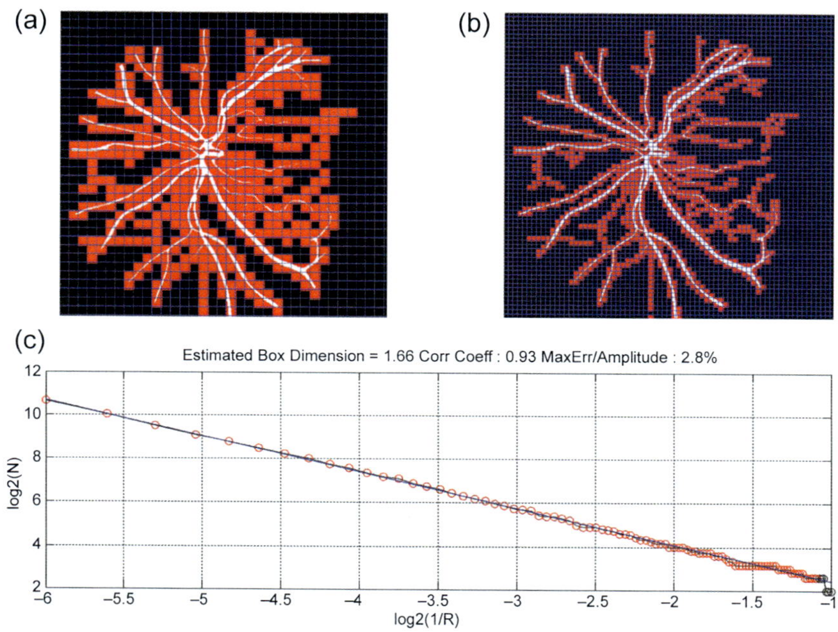

Figure 7.2. Mesh grid of varying box-sizes (scales) covering binerized retinal vasculature. (a) e.g., Box size of R = 8 Pixels (b) e.g., Box size of R = 16 Pixels (c) The slope of the straight line fitted on *Log* (*N*) vs. *Log* (1/*R*) is the box-counting fractal dimension (Here: 1.66).

7.5 Differential (3D) Box-counting Dimension

Differential Box-Counting [6] is a modification of the binary Box-Counting introduced in section 7.4, with the advantage of being applicable to gray scale images which does not require image segmentation or binarization. The method is based on the basic binary Box-counting given by eqn. 7.4 but with different manner of counting the number of boxes (N). Assuming an image of size $S \times S$ is a scaled version of its original size $M \times M$ pixels with S being an integer between 1 and $\frac{M}{2}$ $(1 < S \leq \frac{M}{2})$. The side length of size R is then estimated as $R = \frac{M}{S}$. The image is considered in a 3D space with (x, y) denoting the 2D coordinates (positions) on a plane and the gray level intensity as the 3rd dimension. The 2D coordinate was covered with grids of size $S \times S$ each having a column of boxes of size $S \times S \times S'$ with S' being height of a single box. Considering total number of gray levels is G (i.e., for 8 bit gray scale image $G = 255$), then $\left\lfloor \frac{G}{S'} \right\rfloor = \left\lfloor \frac{M}{S} \right\rfloor$ where $\lfloor x \rfloor = floor\,(x)$ denoting the largest integer not greater than x. The boxes are numbered sequentially as shown in Fig. 7.3.

For each grid (i, j) and using the numbers assigned to each box, the distance between the boxes containing minimum and maximum gray level of the image (n) is obtained as follows to find the contribution of N in the $(i, j)^{th}$ grid:

$$n(i,\ j) = l - k + 1 \tag{7.5}$$

where l and k are the numbers associated with the boxes containing maximum and minimum gray levels respectively. The overall N corresponding to each value of R (i.e., each value of S) can be obtained by adding all the individual contributions:

$$N = \sum_{i,j} n(i, j) \tag{7.6}$$

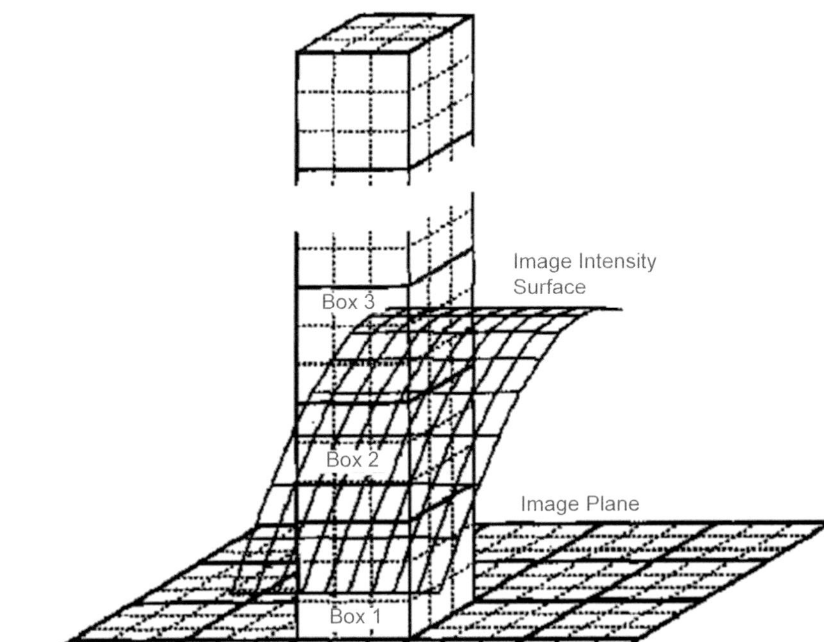

Figure 7.3. Illustration of Gray scale (differential) box counting [6] (© 1994 IEEE with some modifications).

Likewise the Binary BC in eqn. 7.4, a *Log-Log* plot of *Log* (*N*) *vs.* $Log(\frac{1}{R})$ yields Gray-scale (differential) BC fractal dimension. This can be interpreted as the self-similarity within the image, and the value of the FD is between 2 and 3.

7.6 Spectral Fractal Dimension

Spectrum FD (SFD) also referred to as Fourier FD (FFD) is another form of FD which is mathematically obtained from the Fourier transform of an image (signal). It is basically calculated from an exponent relation between the power spectrum of the Fourier transform of the image and the frequency variable [13]. The option of computing fractal dimension of the image in the spectral domain provides number of benefits. FFD is invariant to geometrical transforms and provides good computational and noise performance compared to other FDs. Assuming $I(x, y)$ a square image of size $N \times N$, the first step to obtain FFD is to calculate the 2D Discrete Fourier Transform (DFT) of the image as in eqn. 7.7.

$$F(p,q) = \frac{1}{N^2} \sum_{n=1}^{N} \sum_{n=1}^{N} I(x,y) e^{-i2\pi(\frac{px}{N}+\frac{qy}{N})} \tag{7.7}$$

where $F(p, q)$ is the DFT coefficients of $I(x, y)$ over finite square region of size $1 \le n \le N$. The magnitude (M) of the DFT (power spectrum) is then obtained at every pixel using the following equation eqn. 7.8.

$$M(p, q) = log\ (\,|\,F(p, q)\,|\,^2) \tag{7.8}$$

where $F(p, q)$ is the magnitude of DFT squared at pixels' location of p and q as in eqn. 7.7.

The following example shows the $M(p, q)$ of a grayscale enhanced retina image (Fig. 7.4). Image enhancement is often performed to ensure that the FFD does not get degraded by the artifacts such as uneven illumination, ocular media opacity, poor contrast and background noise. The enhancement is usually performed by wavelet transform and application of a 2D Gabor wavelet as the kernel of a directional matched filter. The matched filter is employed over a number of directions starting from 0 to 170° with 10° intervals. The responses are then maximized at every pixel and over each direction. The complete image enhancement process has been described previously by Soares et al. [14].

In this example the zero-frequency (DC) component of the DFT has been shifted from the corners to the center of the spectrum also known as the "origin", for better visualization. This will result into the following transformed image as shown in Fig. 7.4.b.

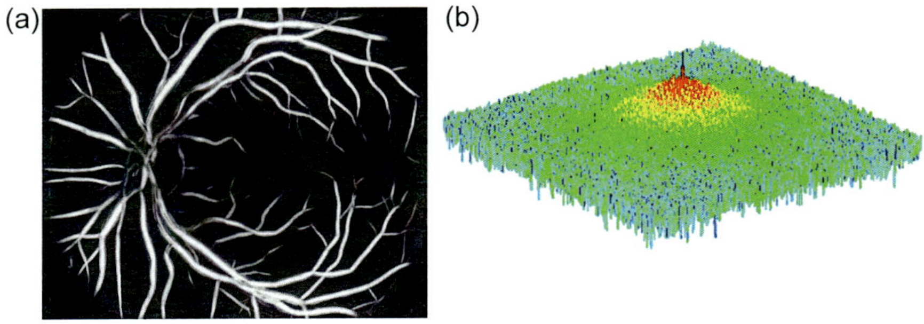

Figure 7.4. DFT of an enhanced retina image (Right eye – Macula centered). (a) Enhanced image using 2D Gabor matched filter in grayscale format. (b) The magnitude of the DFT of the enhanced image with zero-frequency (DC) component shifted to the center of the array (Presented as 3D color map).

When an image is Fourier Transformed (assuming the image to be a fractal surface), $M(p, q)$ will be linearly related to the log of frequency. To obtain FFD, $M(p, q)$ has to be sampled radially from the center of the image towards the peripheral region on a circle with radius of ω, where ω is the frequency which increases with the Euclidian distance from the origin.

To calculate FFD, the $M(p, q)$ is plotted versus log of frequency and the slope of the plot (β) corresponding to the slope of the best fitting line is obtained. β will always be negative as the values of $M(p, q)$ decline with increase in the log of frequency (in Fig. 7.5.c). The FFD is related to β by the following formula:

$$FFD = \frac{\beta + 6}{2} \tag{7.9}$$

However, there will be some variations in spectral properties along different directions referred to as the anisotropic behavior of the image which will result into profiles with different slopes (i.e., different βs with large SD). The rose plot (Fig. 7.5.b) is therefore a convenient method to use for the quantitative determination of the anisotropic behavior of textures with self-similar properties. It shows the slope of the profile as a function of orientation and is used to check whether the surface is isotropic with same dimensions in all directions. In practice for computational purpose the profiles are averaged and the slope of the resulting profile is used in eqn. 7.9 to obtain the FFD.

7.7 Higuchi's Fractal Dimension

Higuchi measures the FD of a set of points in the form of 1D time series [8]. It can address some limitations of some fractal algorithms such as the box counting, and FFD through obtaining the FD along a specific direction

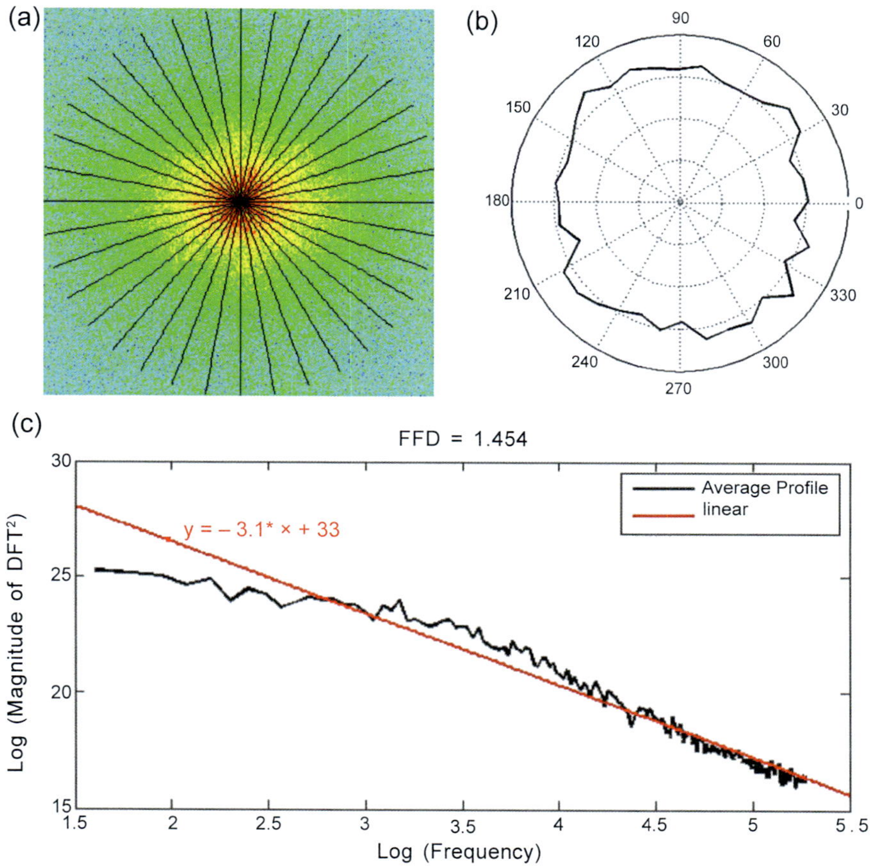

Figure 7.5. Calculation of FFD. (a) 2D representation of Magnitude of the power spectrum, the black radial lines show the sampling direction (here spaced at 5° interval for better presentation) (b) The rose plot of various βs corresponding to the slope of the profiles sampled along different directions. (c) plot of average $M(p, q)$ vs. Log (Frequency) and the best fitting line to the data (slope $\beta = -3.1$).

on the image plane. In order to find FD of images (as 2D objects), image intensities need to be transformed into series of 1D signals. This is possible by scanning the image along different paths including horizontal and vertical in such a way as to cover the entire Region of Interest (ROI). This means the image of size (M × N) needs to be decomposed into its M rows and N columns for extraction of the intensity values [15]. In the horizontal and vertical scanning, the gray values are scanned along horizontal and vertical lines and a set of $FD_{H(1\text{-}M)}$ and $FD_{V(1\text{-}N)}$ values corresponding to the number of columns and rows is obtained. The overall FD_H and FD_V can be obtained by averaging the horizontal ($FD_{H(1\text{-}M)}$) and vertical ($FD_{V(1\text{-}N)}$) dimensions respectively.

Assuming $s(1), s(2),\ldots, s(N)$ is a set of N discrete gray values (intensities) recorded along each scanning path also called image profile, K subsets of new data series $S_k^1, S_k^2, \ldots, S_k^k$ are defined as follow:

$$S_k^m: \quad s(m), s(m+k), s(m+2k),\ldots, s\left(m+\left[\frac{N-m}{k}\right].k\right) \quad (m=1,2,\ldots,k) \tag{7.10}$$

Where [ ] is the Gauss' notation representing the *floor* function (the largest previous integer). m and k are the integers denoting the initial and the intervals respectively. Parameter $L_m(k)$ is calculated as in (2) for each S_k^m corresponding to the curve length of that subset.

$$L_m(k) = \frac{A}{k}\left(\sum_{i=1}^{\left[\frac{N-m}{k}\right]} \left|s(m+ik) - s(m+(i-1)k)\right|\right) \tag{7.11}$$

where $A = \dfrac{N-1}{k\left[\dfrac{N-m}{k}\right]}$ is the normalizing factor for the subset curve length.

The total curve length is obtained by averaging the subset lengths over the whole interval (k sets of $L_m(k)$). The maximum value that k can obtain is equal or less than $\left[\dfrac{N}{2}\right]$. The change of this length as a function of the interval, on the logarithmic scale, is considered as measure of Higuchi's FD.

Depending on the irregularity of the geometry and structure of the object on the image, other scanning directions such as radial, diagonal and circular can also be considered. Selection of an appropriate scanning direction depends on the purpose of study as whether it is to measure the FD corresponding to the entire image or cross-section profile of an object of interest on the image. For instance, if the geometry of interest is retinal vasculature, one possible option for calculating Higuchi's FD is to scan the image along concentric circles centered at ODC with the circles spaced with one pixel intervals and with the radii of the innermost circle greater than the OD radius. This will result in a set of $FD_{c\text{-}i}$ with the i being the total number of circles obtained. The overall FD_c is calculated by averaging the $FD_{c\text{-}i}$ values. The choice of circular scanning path for retinal images is due to the radial orientation of the major vessels (arteries and veins). Such concentric paths will cut through the major vessels resulting into a series of data profiles including vessel diameter information. It should be noted that, although Higuchi is a 1D method, FD values range between 1 and 2 as the time series signals are embedded in a 2D plane [8] similar to the box-counting method.

7.8 Summary

The chapter has discussed the concept of self-similarity, entropy and fractal dimension and association of each with the degree of disorder, randomness, irregularity and even complexity of natural structures. These measurements have the advantage of summarizing the entire image to a single number, making it suitable for quantitative analysis. In this chapter, the concept of tortuosity was also introduced to measure the degree of curvature of a tubular structure such as retinal vessel. Image of retina and the vascular trees were taken as an example of medical image including natural structures with self-similar and fractal-like properties, to explore a number of FD methods frequently used in medical image analysis (i.e., Binary Box-counting, Differential 3D Box-counting, Fourier (Spectral) Fractal and Higuchi's method).

References

1. Azemin, M.Z.C., D.K. Kumar, T.Y. Wong, J.J. Wang, P. Mitchell, R. Kawasaki and H. Wu. 2012. Age-related rarefaction in the fractal dimension of retinal vessel. Neurobiology of Aging, 33: 194.e1–194.e4.
2. Che Azemin, M.Z., D.K. Kumar, T.Y. Wong, R. Kawasaki, P. Mitchell and J.J. Wang. 2011. Robust methodology for fractal analysis of the retinal vasculature. Medical Imaging, IEEE Transactions on, 30: 243–250.
3. Kawasaki, R., M.Z. Che Azemin, D.K. Kumar, A.G. Tan, G. Liew, T.Y. Wong, P. Mitchell and J.J. Wang. 2011. Fractal dimension of the retinal vasculature and risk of stroke: a nested case-control study. Neurology, 76: 1766–7.
4. Mandelbrot, B.B. 1983. The Fractal Geometry of Nature, Henry Holt and Company.
5. Jürgens, H., D. Saupe and H.O. Peitgen. 1992. Fractals for the Classroom: Part One Introduction to Fractals and Chaos, Springer.
6. Sarkar, N. and B.B. Chaudhuri. 1994. An efficient differential box-counting approach to compute fractal dimension of image. Systems, Man and Cybernetics, IEEE Transactions on, 24: 115–120.
7. Azemin, M.Z., D.K. Kumar, T.Y. Wong, R. Kawasaki, P. Mitchell and J.J. Wang. 2011. Robust methodology for fractal analysis of the retinal vasculature. IEEE Trans. Med. Imaging, 30: 243–50.
8. Higuchi, T. 1998. Approach to an irregular time series on the basis of the fractal theory. Physica D: Nonlinear Phenomena, 31: 277–283.
9. Raghavendra, B.S. and D. Narayana Dutt. 2010. Computing fractal dimension of signals using multiresolution box-counting method. World Acad. Sci. Eng. Technol., 6: 1223–1238.
10. Masters, B.R. 2004. Fractal analysis of the vascular tree in the human retina. Annu. Rev. Biomed. Eng., 6: 427–52.
11. Family, F., B.R. Masters and D.E. Platt. 1989. Fractal pattern formation in human retinal vessels. Physica D: Nonlinear Phenomena, 38: 98–103.
12. Liew, G., J.J. Wang, P. Mitchell and T.Y. Wong. 2008. Retinal vascular imaging: a new tool in microvascular disease research. Circ. Cardiovasc. Imaging, 1: 156–61.
13. Russ, J.C. 1994. Fractal Surfaces, Plenum Press, New York.
14. Soares, J.V., J.J. Leandro, R.M. Cesar Junior, H.F. Jelinek and M.J. Cree. 2006. Retinal vessel segmentation using the 2-D Gabor wavelet and supervised classification. IEEE Trans. Med. Imaging, 25: 1214–22.
15. Ahammer, H. 2011. Higuchi Dimension of Digital Images. PLoS ONE, 6: e24796.

CHAPTER 8

Fractal Dimension of Retinal Vasculature

ABSTRACT

This chapter will introduce the reader to the application of fractal dimension (FD) of retinal images in analysis of retinal vasculature pattern, retinopathies and disease risk assessment. Advantages of using FD are that it summarizes the information into a single value which is very convenient in terms of feature reduction and for statistical analysis. It is also suitable for automatic assessments without requiring extensive user intervention.

This chapter covers a brief description of human eye anatomy with emphasis on the retina structure and abnormalities due to systemic diseases which manifest themselves in retina. The link between such abnormalities, change in retinal vessel morphology and the FD are then introduced and discussed.

8.1 Introduction to Human Eye Anatomy

A cross-sectional view of the human eye is shown in Fig. 8.1. It consists of three major layers including external, intermediate and internal layers. The external layer includes sclera (white of the eye) and cornea which forms the supporting wall of the eyeball. The cornea is a transparent layer covering both iris and pupil. Iris is the coloured circular muscle which controls the size of the pupil (a dark circular opening in the iris) and the amount of light entering the eye ball based on the environmental lighting conditions. The second layer is the intermediate layer which is subdivided into two anterior (iris and ciliary body) and posterior (choroid) parts. The internal layer is the retina (Fig. 8.1), a thin (thickness: 0.5 mm approx.) multi-layer photon sensitive circular disc of diameter between 30 and 40 mm [1]. In the centre of retina there is a circular to oval shape like area of about 2 × 1.5 mm called the optic nerve. The blood vessels nourishing the retina enter/exit the eyeball through the centre of the optic nerve. Optic nerve also carries the electrical impulses to the visual cortex of the brain which are generated by the light-sensing cells on the retina. There is small

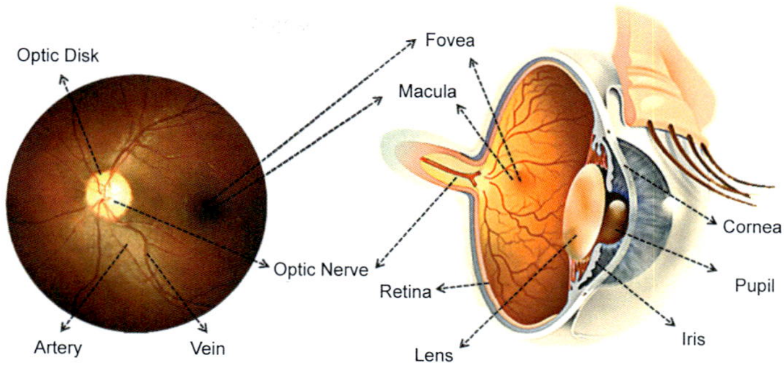

Figure 8.1. Major structures in human retina. (a) Cross-sectional illustration of human eye [Source: WebMD LLC [2]]. (b) Human retina photograph taken by fundus camera (Cannon CR-1).

extra sensitive area located in the centre of the retina and is responsible for central vision. This area is called Macula and contains fovea in the middle that provides clearest vision.

There are two sources of blood supply nourishing the retina. The choroidal blood vessels that carry about 65 to 85% [3] of blood flow to the outer retina, particularly to the photoreceptor cells and the central retinal artery that supply inner retinal layers from the optic nerve head. Within the optic nerve head the artery subdivides in to four main branches to supply the four quadrants of the retina. The large arteries in each quadrant extend within the retina toward the periphery areas and are progressively subdivided to form arteries with smaller diameters. When they reach ora serrata (a junction between retina and ciliary body), they return as a continuous venous drainage system with distribution pattern similar to the arteries. In general these vessels emerge in radial direction from the optic nerve and curve towards and around the fovea.

8.2 Eye Fundus Retinopathy–Disease Manifestation in Retina

Retina manifests a number of disease conditions. Hence it provides a unique opportunity for screening of abnormalities and disease risk assessment. This uniqueness is mainly due to the properties of retinal microcirculation which share similar anatomical, physiological, and embryological characteristics to cerebral vessels [4] and also the ability of retina to be viewed non-invasively and *in vivo* with the help of fundus photography. Examples of the ocular (eye-related) diseases include Macular Degeneration (MD) and glaucoma. MD, often referred to as Age-related Macular Degeneration (AMD), is a type of degenerative disease that causes progressive loss of central vision. Vision loss occurs because of breakdown or thinning of a layer of cells underneath the retina (Retinal Pigment Epithelium (RPE)) building up fatty deposits (drusens) in the macular region (dry AMD) or due to abnormal growth of choroidal vascular structure into the macula (wet AMD) [5]. Glaucoma is characterized by gradual damage to the optic nerve which leads to permanent vision loss and blindness if it is not detected and treated early. Glaucoma manifests in retina by cupping of the optic nerve head. Therefore the ratio of the optic disc cup and the neuroretinal rim surface area (area of optic disk containing neural elements) as well as its color and uniformity, can be indicative of the presence and progression of glaucoma.

Along with ocular diseases, retina can also be affected by a number of systemic diseases which manifest themselves with certain retinopathy signs. Diabetes mellitus (DM) and cardiovascular diseases (CD) are two examples of the most prevalent systemic diseases. Their ocular manifestation can be viewed and studied *in-vivo* via eye fundus photography and retinal image analysis.

8.3 FD and Age Related Changes of Retinal Vasculature

Research has shown that there is degradation in anatomical structure and complexity of a variety of organ systems with age. Examples includes loss of dendritic arbor in aging cortical neurons [6], reduction in complexity of cardiac activity [7], electroencephalography (EEG) [8] as well as retinal vasculature [9]. Therefore fractal theory can be applied to retinal photographic images to quantify the association between aging and the change in complexity of retinal vessels. It can also be used as a baseline measure for diagnostic and prognostic applications in future. This section describes the experiments and results obtained while studying this association, and discuss the results in relation to the observed reduction in fractal properties of retinal vasculature.

Recent works suggest inverse correlation of binary box-counting FD with age; validated using three hundred disk-centred retinal photographs [10]. However, due to the need for image segmentation which can be degraded by the background noise; measurement of FD using binerized box-counting method may result in imprecise FD estimates. This may be the reason why some studies have been unsuccessful in observing loss of complexity in retinal vasculature as a result of aging [11,12]. In order to overcome this problem, Azemin et al. suggested the use of FFD together with gray-scale enhanced retinal image as a three-dimensional (3D) representation of the vasculature [13].

In this work, close to 400 age-stratified sample of healthy subjects (i.e., 50% men and 50% women) were randomly selected from the Blue Mountain Eye Study (BMES) dataset [14]; a population-based study conducted in a suburban region west of Sydney, Australia. The age range was subdivided into four narrow groups of 50–59, 60–69, 70–79 and 80–89 years, consisting of retinal images from the both eyes. FFD was computed using the technique proposed by Russ et al. [15] and in a circular region of interest with the radius equivalent to 150 pixels from the center of the optic disk.

The result showed a linear decline in FFD with increase in age and a statistically significant ($P < 0.0001$) association with this factor. This was regardless of the type of images (i.e., left or right) and gender of the subjects used. Results from linear regression analysis also revealed 1% decline in FFD with every increase in decade of age. The findings imply that the age factor will have a confounding effect in any case control studies when assessing the association of FD with disease of interest. However, limitation of this study is that it only covers a narrow age spectrum in which retinal images of the people older than 49-year have been included.

8.4 FD and Hypertensive Retinopathy

Fundus photography also allows for risk assessment of cardiovascular diseases and its subsequent complications. This includes hypertensive retinopathy which is referred to retina damage due to hypertension (high blood pressure). It is associated with long term risk of stroke [16,17] as the second commonest cause of death worldwide [18]. Some of the first observable signs are flame hemorrhages and cotton wool spots. As the disease progresses, hard exudates may appear around the macula along with swelling of the macula and the optic nerve resulting into vision impairment. Another, associated clinical biomarker is reduction in arteriolar diameter which is assessed by measuring the reduction in arteriolar diameter to venous diameter ratio (AVR) [19,20]. However, reduced AVR does not necessarily indicate constant venular diameter and therefore reflecting generalized arteriolar narrowing. Venular diameters may also change (usually increase) due to different pathologic conditions including atherosclerosis, inflammation, cholesterol levels and stroke event [21]. For instance, this ratio for the stroke incidence corresponds to both arteriolar narrowing and venular widening [5,22]. Therefore using AVR, the exact contribution of individual arteriolar and venular diameters to disease event may remain unknown. For more effective translation of retinopathy signs into clinical application and assessment purpose, hypertensive retinopathy, in a relatively new classification, is classified into four levels of none, mild, moderate and severe [23] based on the observable retinopathy signs. Generalized and focal arteriolar narrowing and arteriovenous nicking are referred to as mild hypertensive retinopathy signs. The presence of lesions such as microaneurysms and hemorrhages, hard and soft exudates (cotton wool spots) indicate the moderate level while optic disc edema is referred

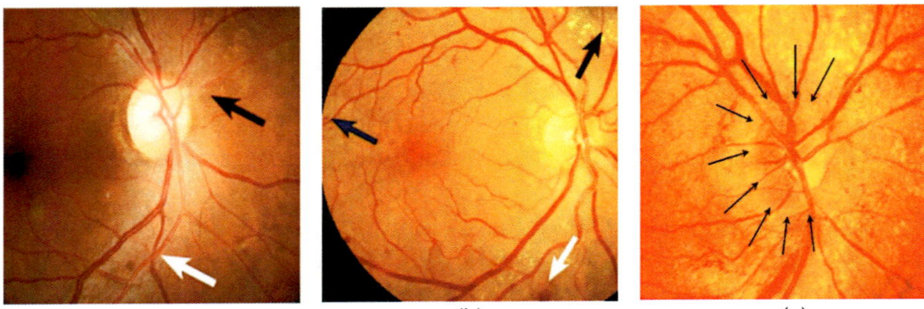

Figure 8.2. Hypertensive retinopathy signs. (a) Mild: generalized and focal arteriolar narrowing (black arrow), arteriovenous nicking (white arrow) [Source: [24]] (b) Moderate: arteriolar narrowing with arteriovenous nicking, flame-shaped hemorrhages (white arrow), hemorrhages (blue arrow), and hard exudates (black arrow) [Source: [24]] (c) Severe: Optic disc edema [Source: [25]].

to as a sign for severe stage [23]. Example of retinal images with some retinopathy signs are shown in Fig. 8.2.

8.5 FD and Risk of Stroke Event

Stroke, also known as "Brain attack", is when brain cells start to die as a result of poor or lack of blood flow to area of the brain. When the brain cells die during a stroke, the affected area in the brain stop functioning properly and cause inability to move, control or feel part of the body which is controlled by that area. There are two main types of stroke, one, as a consequence of occlusion of a blood vessel supplying the brain (i.e., Ischemic stroke) and the second, due to burst of brain aneurism and leakage from a weakened blood vessel (i.e., hemorrhagic stroke).

Studies have shown that changes in retinal vasculature can be associated with small infarcts observable from brain imaging [26] as retinal and cerebrovascular network share similar characteristics. This association appears as some changes in retinal vessel caliber and in overall complexity of geometrical pattern of the vasculature. This change of complexity can be summarized and quantified by Fractal Dimension (FD) [26,27] as a global measure which has been used in a number of studies for stroke risk assessment [9,26,28,29].

Cheung et al. investigated the association between FD of retinal images and different ischemic stroke subtypes including lacunar (i.e., occlusion of a single small arterial vessel supplying the brain's deep structures), large artery and cardioembolic strokes. In a cohort comprising 392 patients with acute ischemic stroke, they found patients with increased box-count FD had higher chance of having lacunar stroke than other stroke subtypes regardless of the age, gender, retinal vessel caliber and, vascular risk factors. However, although the box-count FD seems to have provided promising results, the technique is susceptible to the measurement method and at risk of being degraded by image artifacts as a result of segmentation or binerization process [29]. Therefore, in order to overcome such limitation, Azemin et al. [9] proposed the Fourier FD (also known as spectrum fractal dimension (SFD)) and applied the technique onto the gray scale (green channel) retinal images after being enhanced by Gabor wavelet transform as explained in Chapter 7. Later application of this work was tested by Kawasaki et al. [26] in a cohort study of stroke risk assessment using the BMES data set. They found a statistical significance ($p = 0.044$, $\alpha = 0.05$) between the case and control groups with lower spectrum fractal dimension (SFD) of 1.504 (95%CI, 1.499–1.510) associated with higher risk of stroke compared to the lower SFD 1.511 (95%CI, 1.507–1.515).

In another study, the association between the change in FD of retinal vessel and future episode of stoke was investigated using a cohort of BMES dataset [30]. In brief, retina images of both eyes were included and the participant's age was limited to a narrow range of 50 to 89 to compensate for the effect of age as possible confounding factor. Total numbers of 104 subjects were examined with confirmed episode of stroke after five-year follow-up, comprising of 21 stroke events, 86 stroke-related deaths, and 3 persons overlapping with both. After adjusting for hypertension as stroke risk factor and diabetes, 46 cases were left for the analysis and matched with another 39 controls based on the age factor (mean (SD) = 67.76 (5.72)), metric body mass index (BMI (Kg/m^2)) (26.32 (4.35)), blood pressure (mmHg) (systolic: 150.64 (18.94), diastolic: 83.40 (10.35)), and history of smoking. Retinal images were enhanced and transformed to gray scale level using the same Gabor wavelet technique as used by Azemin et al. [9]. Higuchi's FD was measured on a series of circular scanning paths placed around the optic disk region, while the radii of the inner and outermost circle was set between one to four optic disk radius from its margin with one pixel interval. The reason behind the choice of concentric circles was that the major vessels in retina are in radial direction with respect to the optic disk location. Such path would intersect with a large number of major vessels and the intensity profiles would include vessel width information. Except Higuchi's FD, spectral and binary box-counting FDs were also calculated and compared. Non-parametric statistical analysis revealed significant association between Higuchi's FD and future episode of stroke ($P = 0.16$) with lower median values for the cases compared to the control group. However, no statistical significance was found for other FD techniques (all P_values > 0.05).

8.6 FD and Diabetic Retinopathy

Diabetes mellitus (DM), simply known as "diabetes" is chronic abnormality of blood glucose metabolism resulting from either altered (type 1 diabetes) or insufficient and ineffective insulin production (type 2 diabetes) by the pancreas. In recent year, there has been a significant increase in the number of diabetic people in the world [31], with a very steep rise in diabetes among the younger cohort [32]. The prevalence is significantly higher in the ethnic groups, and often in countries with lower availability of quality healthcare facilities. Untreated diabetes can lead to a number of complications, such as diabetic retinopathy (DR) [33], and diabetic neuropathy [34]. Diabetic patients are more likely to suffer from blindness, ventricular arrhythmia, silent ischaemia, sudden cardiac death and stroke compared with other people. DM will also damage the vessel walls as a type of retinopathy. The corresponding effects on retina include emergence of exudates, haemorrhages, microaneurysms, re-growth of new blood vessels (Neovascularisation) and cotton-wool patches (retinal infarcts) some of which have been shown in Fig. 8.3.

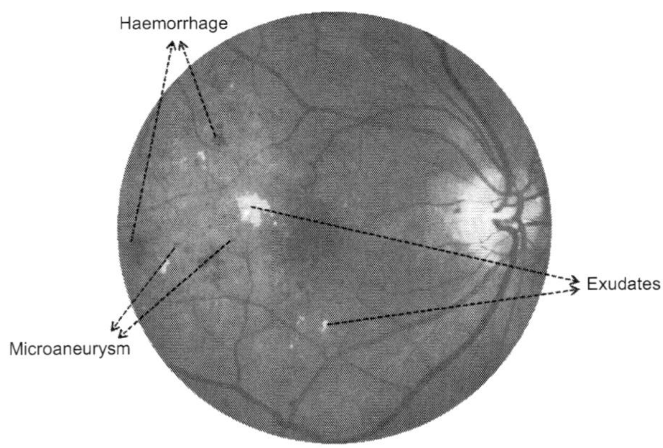

Figure 8.3. Example of a fundus image with severe diabetic retinopathies [Source: [1] with some modifications].

There have been successes with the population screening to identify diabetes [35–38]. However, opportunistic evaluation accounts for the largest detection of diabetic patients among the low risk population and is largely based on the visit of the person to their primary health provider for other health factors. With the reduction in the age of the diabetic patients [39,40] commonly appearing among teenagers, a number of patients either go undiagnosed for a significant period [41] or get diagnosed subsequent to the manifestation of the secondary symptoms.

Studies have assessed the association between a number of disease biomarkers in retina and risk of diabetes which includes the change in retinal vascular diameter [42], tortuosity and early pathological alterations. It has also been used as a predicting factor for retinopathy risk assessment in terms of whether the patients may develop retinopathy and other complications in future [42]. For middle-aged persons (45 to 64), retinal arteriolar narrowing is associated with incidence of diabetes independent of the known risk factors [43]. Elsewhere, significantly wider summary retinal arteriolar and venular caliber has been reported in patients aged 45 to 84 with mild and no diabetic retinopathy [20,44] and used for analysis of the progression of retinopathy in type 1 cases [45,46]. However, vessel diameter is one of the many structural parameters of retinal vasculature which conveys a very specific aspect of the retinal vascular network. Other structural changes including larger arteriolar branching angle and increased tortuosity have also been reported as indicator of longer duration of diabetes and higher A1C respectively [47]. Diversity of such parameters and a lack of method to summarize geometrical variations of branching pattern have led to consideration of FD for quantifying early diabetic microvascular damage. Such measure has shown promising outcome in microvascular disease assessment and is highly correlated with a number of biological parameters [48]. The study of early diabetic retinopathy shows that increase in retinal vascular FD is significantly associated with increasing odds of retinopathy [49]. In patients with type 1 diabetes, the study found that persons with lower FD were more likely to have proliferative retinopathy [50]. No association between FD and any retinal circulatory parameters of the retinal arterioles was found in patients with type 2 diabetes mellitus [51].

In a more recent work, the association between mild Non-proliferative diabetic retinopathy (NPDR) and differential (3D) box-counting (DBC) has been reposted using retinal images from an Indian population with type 2 diabetes [52]. A total number of 189 disk centred retinal images were used for the analysis comprising of 23 diabetic and 166 non-diabetic matching cases. Among all 23 diabetic cases, 5 participants had very mild non-proliferative diabetic retinopathy (NPDR), as identified by the presence of observable damaged blood vessels and micro-aneurism while the retina scan of the remaining 18 did not show any observable retinopathy. Participants were labelled as diabetic based on fasting and post-prandial glucose level and were checked to make sure they do not have history of any other systemic disease including stroke, hypertension and cardiovascular disorder. The diabetic subjects were characterized based on age (mean $\pm$ SD $= 52.69 \pm 7.90$), metric body mass index (BMI (Kg/m2)) (27.60 ± 4.56), blood pressure (mmHg) (systolic: 143.04 ± 19.52, diastolic: 81.73 ± 11.83) and gender (54% female).

Analysis of variance (ANOVA) test showed a significant change in DBC FD of the eye fundus images with early stage of diabetic retinopathy (DR). The result also confirmed DBC FD of retinal images tend to be higher ($p < 0.001$) in diabetic patients (2.440, CI 95% 2.433–2.447) compared to healthy subjects (2.422, CI 95% 2.420–2.425) showing an increase in the overall complexity of retinal vessel as a result of diabetes (i.e., NPDR or no DR). This indicates that DBC can be potentially used to quantify changes in the complexity of retinal vasculature even ahead of DR.

Also multivariate regression analysis was used to study the association of BMI, age, blood pressure (systolic and diastolic) and gender, as potential confounders. Two different models were constructed to determine the effect of limiting the age factor. In the first model (model 1), FDs corresponding to the entire population (a wide age range of 14 to 73) were modelled by the above predictors plus the diabetes factor. In this model, age, gender and diabetes were found to be significantly associated with FD variation (all p values < 0.05).

The second model (model 2) was constructed similar to model 1 except for the age range which was bounded between 50 to 73 years. This new range was chosen based on prevalence of diabetes which was higher among the older population and also to remove the skewness from the data due to age factor. In this model and unlike model 1, the age factor was not found to be associated with FD ($p = 0.79$). However, FD showed significant association with gender and diabetes in both models (all p values < 0.05).

Advantage of this technique is that unlike the binary box-counting, DBC can be applied to enhanced (3D) grayscale images and does not require image segmentation which makes the technique free from segmentation errors and suitable for automated and unsupervised analyses. However, the weakness of this study are the small number of small sample size and consideration of gender as the only confounding factor out of many other possible factors such as age, hypertension, BMI and dyslipidemia. Also inclusion of a wide age spectrum might have led to significant difference between the mean age of the two population groups.

8.7 Summary

Retina provides a unique opportunity and a window for screening of ocular and systemic diseases in human body. It is the only place where the blood vessels are non-invasively accessible through fundus imaging. Diseases such as hypertension, diabetes, stroke as well as natural aging process are known to manifest themselves in retina with certain retinopathy signs affecting the morphology and complexity of the vascular network. Therefore FD has been employed for characterization of these abnormalities associated with different stages of such diseases. This chapter has discussed the eye fundus retinopathy, disease manifestation in retina and the resulting changes to retina followed by a comprehensive review of the FD-based techniques to explore geometrical variations of branching pattern as a result of disease incidence. The changes in retinal FD were found to be associated with different ischemic stroke subtypes, non-proliferative and proliferative diabetes mellitus, hypertensive retinopathy and the ageing.

References

1. Kolb, H., E. Fernandez and R. Nelson. 1995. Simple Anatomy of the Retina—Webvision: The Organization of the Retina and Visual System, Salt Lake City (UT): University of Utah Health Sciences Center.
2. WebMD. 2009. Picture of the eye, ed: WebMD, LLC.
3. Henkind, P., R.I. Hansen and J. Szalay. 1979. Ocular circulation. pp. 98–155. *In*: Records, R.E. (ed.). Physiology of The Human Eye and Visual System, New York: Harper & Row.
4. Wong, T.Y. 2004. Is retinal photography useful in the measurement of stroke risk?, The Lancet Neurology, 3: 179–183.
5. Abramoff, M.D., M.K. Garvin and M. Sonka. 2010. Retinal imaging and image analysis. IEEE Trans. Med. Imaging, 3: 169–208.
6. Goldberger, A.L., L.A. Amaral, J.M. Hausdorff, P. Ivanov, C.K. Peng and H.E. Stanley. 2002. Fractal dynamics in physiology: alterations with disease and aging. Proc. Natl. Acad. Sci. U S A, 99(Suppl 1): 2466–72.
7. Pikkujamsa, S.M., T.H. Makikallio, L.B. Sourander, I.J. Raiha, P. Puukka, J. Skytta, C.K. Peng, A.L. Goldberger and H.V. Hurikuri. 1999. Cardiac interbeat interval dynamics from childhood to senescence: comparison of conventional and new measures based on fractals and chaos theory. Circulation, 100: 393–9.
8. Kaplan, D.T., M.I. Furman, S.M. Pincus, S.M. Ryan, L.A. Lipsitz and A.L. Goldberger. 1991. Aging and the complexity of cardiovascular dynamics. Biophys J., 59: 945–9.
9. Azemin, M.Z.C., D.K. Kumar, T.Y. Wong, R. Kawasaki, P. Mitchell and J.J. Wang. 2011. Robust methodology for fractal analysis of the retinal vasculature. Medical Imaging, IEEE Transactions on, 30: 243–250.
10. Liew, G., J.J. Wang, N. Cheung, Y.P. Zhang, W. Hsu, M.L. Lee, P. Mitchell, G. Tikellis, B. Taylor and T.Y. Wong. 2008. The retinal vasculature as a fractal: methodology, reliability, and relationship to blood pressure. Ophthalmology, 115: 1951–6.
11. Family, F., B.R. Masters and D.E. Platt. 1989. Fractal pattern formation in human retinal vessels. Physica D: Nonlinear Phenomena, 38: 98–103, 1989/09/01.
12. Masters, B.R. 2004. Fractal analysis of the vascular tree in the human retina. Annu. Rev. Biomed. Eng., 6: 427–52.
13. Azemin, M.Z., D.K. Kumar, T.Y. Wong, J.J. Wang, P. Mitchell and R. Kawasaki. 2012. Age-related rarefaction in the fractal dimension of retinal vessel. Neurobiol. Aging, 33: 194 e1–4.
14. Mitchell, P., W. Smith, K. Attebo and J.J. Wang. 1995. Prevalence of age-related maculopathy in Australia. The Blue Mountains Eye Study. Ophthalmology, 102: 1450–60.
15. Russ, J.C. 1994. Fractal Surfaces, Springer US.
16. Ong, Y.T., T.Y. Wong, R. Klein, B.E. Klein, P. Mitchell and A.R. Sharrett. 2013. Hypertensive retinopathy and risk of stroke. Hypertension, 62: 706–11.
17. Gorelick, P.B. 2002. New horizons for stroke prevention: PROGRESS and HOPE. The Lancet Neurology, 1: 149–156.
18. Toronto, J.W.N.P.N.U. and V.H.P.N.U.W. Ontario. 2001. Stroke Prevention, Oxford University Press, USA.
19. Wong, T.Y., M.D. Knudtson, R. Klein, B.E.K. Klein and L.D. Hubbard. 2004. A prospective cohort study of retinal arteriolar narrowing and mortality. American Journal of Epidemiology, 159: 819–825.
20. Wong, T.Y., F.M.A. Islam, R. Klein, B.E.K. Klein, M.F. Cotch, C. Castro, A.R. Sharrett and E. Shahar. 2006. Retinal vascular caliber, cardiovascular risk factors, and inflammation: The Multi-Ethnic Study of Atherosclerosis (MESA). Investigative Ophthalmology & Visual Science, 47: 2341–2350.

21. Ikram, M.K., F.J. de Jong, J.R. Vingerling, J.C.M. Witteman, A. Hofman, M.M.B. Breteler and P.T. deJong. 2004. Are retinal arteriolar or venular diameters associated with markers for cardiovascular disorders? The Rotterdam Study. Investigative Ophthalmology & Visual Science, 45: 2129–2134.
22. McGeechan, K., G. Liew, P. Macaskill, L. Irwig, R. Klein, B.E.K. Klein, J.J. Wang, P. Mitchell, J.R. Vingerling, P.T. deJong, J.C. Witteman, M.M. Breteler, J. Shaw, P. Zimmet and T.Y. Wong. 2009. Prediction of incident stroke events based on retinal vessel caliber: A systematic review and individual-participant meta-analysis. American Journal of Epidemiology, 170: 1323–1332.
23. Liew, G. and J.J. Wang. 2011. Retinal vascular signs: A window to the Heart?, Revista Española de Cardiología (English Version), 64: 515–521.
24. Bhargava, M. and T.Y. Wong. 2006. Current concepts in hypertensive retinopathy. Retinal Physician, 10: 43–54.
25. The University of Iowa. Optic disc edema. Available: http://webeye.ophth.uiowa.edu/dept/coms/grading/optic-disc-edema.htm.
26. Kawasaki, R., M.Z. Che Azemin, D.K. Kumar, A.G. Tan, G. Liew and T.Y. Wong. 2011. Fractal dimension of the retinal vasculature and risk of stroke: A nested case-control study. Neurology, 76: 1766–1767.
27. Che Azemin, M.Z., D.K. Kumar, T.Y. Wong, J.J. Wang, R. Kawasaki and P. Mitchell. 2010. Retinal stroke prediction using logistic-based fusion of multiscale fractal analysis, in Imaging Systems and Techniques (IST), 2010 IEEE International Conference on, pp. 125–128.
28. Landini, G., G.P. Misson and P.I. Murray. 1993. Fractal analysis of the normal human retinal fluorescein angiogram. Curr. Eye Res., 12: 23–7.
29. de Mendonca, M.B., C.A. de Amorim Garcia, A. Nogueira Rde, M.A. Gomes, M.M. Valenca and F. Orefice. 2007. Fractal analysis of retinal vascular tree: segmentation and estimation methods. Arq. Bras. Oftalmol., 70: 413–22.
30. Aliahmad, B., D.K. Kumar, H. Hao, P. Unnikrishnan, M.Z. Che Azemin, R. Kawasaki and P. Mitchell. 2014. Zone specific fractal dimension of retinal images as predictor of stroke incidence. The Scientific World Journal, 2014, p. 7.
31. CDC. 2005. Centre for disease control, National diabetic fact sheet, USA.
32. Engelgau, M.M., L.S. Geiss, J.B. Saaddine, J.P. Boyle, S.M. Benjamin, E.W. Gregg, E.F. Tierney, N. Rios-Burrows, A.H. Mokdad, E.S. Ford, G. Imperatore and K.M. Narayan. 2004. The evolving diabetes burden in the United States. Ann. Intern. Med., 140: 945–50.
33. Kempen, J.H., B.J. O'Colmain, M.C. Leske, S.M. Haffner, R. Klein, S.E. Moss, R. Hugh, H.R. Taylor, R.F. Hamman, S.K. West, J.J. Wang, N.G. Congdon and Friedman. 2004. The prevalence of diabetic retinopathy among adults in the United States. Arch. Ophthalmol., 122: 552–63.
34. Wong, M.-c., J.W.Y. Chung and T.K.S. Wong. 2007. Effects of treatments for symptoms of painful diabetic neuropathy: systematic review. BMJ, 335: 87, 2007-07-12 00:00:00.
35. Abramoff, M.D., J.M. Reinhardt, S.R. Russell, J.C. Folk, V.B. Mahajan and M. Niemeijer. 2010. Automated early detection of diabetic retinopathy. Ophthalmology, 117: 1147–54.
36. Nguyen, T.T., J.J. Wang, A.R. Sharrett, F.M.A. Islam, R. Klein and B.E.K. Klein. 2008. Relationship of retinal vascular caliber with diabetes and retinopathy: The Multi-Ethnic Study of Atherosclerosis (MESA). Diabetes Care, 31: 544–549.
37. Klein, R., B.E. Klein, S.E. Moss and T.Y. Wong. 2006. The relationship of retinopathy in persons without diabetes to the 15-year incidence of diabetes and hypertension: Beaver Dam Eye Study. Trans. Am. Ophthalmol. Soc., 104: 98–107.
38. Wong, T.Y., Q. Mohamed, R. Klein and D.J. Couper. 2006. Do retinopathy signs in non-diabetic individuals predict the subsequent risk of diabetes?, Br. J. Ophthalmol., 90: 301–3.
39. Rosenbloom, A.L., J.R. Joe, R.S. Young and W.E. Winter. 1999. Emerging epidemic of type 2 diabetes in youth. Diabetes Care, 22: 345–54.

40. Dabelea, D., R.L. Hanson, P.H. Bennett, J. Roumain, W.C. Knowler and D.J. Pettitt. 1998. Increasing prevalence of Type II diabetes in American Indian children. Diabetologia, 41: 904–10.
41. WHO. 2003. World Health Organization Screening for Type 2 Diabetes: Report of a World Health Organisation and International Diabetes Federation Meeting.
42. Ikram, M.K., C.Y. Cheung, M. Lorenzi, R. Klein, T.L.Z. Jones and T.Y. Wong. 2013. Retinal vascular caliber as a biomarker for diabetes microvascular complications. Diabetes Care, 36: 750–759.
43. Wong, T.Y., R. Klein, A.R. Sharrett, M.I. Schmidt, J.S. Pankow, D.J. Couper, B.E. Klein, L.D. Hubbard and B.B. Duncan. 2002. Retinal arteriolar narrowing and risk of diabetes mellitus in middle-aged persons. JAMA, 287: 2528–33.
44. Kifley, A., J.J. Wang, S. Cugati, T.Y. Wong and P. Mitchell. 2006. Retinal vascular caliber, diabetes, and retinopathy. American Journal of Ophthalmology, 143: 1024–1026.
45. Wong, T.Y., A. Shankar, R. Klein and B.E.K. Klein. 2004. Retinal vessel diameters and the incidence of gross proteinuria and renal insufficiency in people with Type 1 diabetes. Diabetes, 53: 179–184.
46. Klein, R., B.E. Klein, S.E. Moss, T.Y. Wong, L. Hubbard, K.J. Cruickshanks and M. Palta. 2004. The relation of retinal vessel caliber to the incidence and progression of diabetic retinopathy: XIX: the Wisconsin epidemiologic study of diabetic retinopathy. Arch. Ophthalmol., 122: 76–83.
47. Sasongko, M.B., J.J. Wang, K.C. Donaghue, N. Cheung, P. Benitez-Aguirre, A. Jenkins, W. Hsu, M.L. Lee and T.Y. Wong. 2010. Alterations in retinal microvascular geometry in young Type 1 diabetes. Diabetes Care, 33: 1331–1336.
48. Cheung, N., K.C. Donaghue, G. Liew, S.L. Rogers, J.J. Wang, S.W. Lim et al. 2009. Quantitative assessment of early diabetic retinopathy using fractal analysis. Diabetes Care, 32: 106–10.
49. Ning Cheung, K., C. Donaghue, G. Liew, S.L. Rogers, J.J. Wang, S.-W. Lim, A.J. Jenkins, W. Hsu, M.L. Lee and T.Y. Wong. 2009. Quantitative assessment of early diabetic retinopathy using fractal analysis. Diabetes Care, 32: 106.
50. Grauslund, J., A. Green, R. Kawasaki, L. Hodgson, A.K. Sjølie and T.Y. Wong. 2010. Retinal vascular fractals and microvascular and macrovascular complications in Type 1 diabetes. Ophthalmology, 117: 1400–1405.
51. Nagaoka, T. and A. Yoshida. 2013. Relationship between retinal fractal dimensions and retinal circulation in patients with Type 2 diabetes mellitus. Current Eye Research, 38: 1148–1152.
52. Aliahmad, B., D.K. Kumar, M.G. Sarossy and R. Jain. 2014. Relationship between diabetes and grayscale fractal dimensions of retinal vasculature in the Indian population. BMC Ophthalmol., 14: 152.

CHAPTER 9

Fractal Dimension of Mammograms

ABSTRACT

Mammograms are the images of the human breast obtained using a low-energy x-ray and are routinely used to examine the breast tissue for presence of any abnormality. These are used to monitor the presence of lumps, masses and cancerous growths for a diagnostic or screening purpose. The contours of breast lesions have fractal properties which are measurable using a range of fractal dimension algorithms.

This chapter will first introduce the breast anatomy and properties of breast tissues observable in digitized mammogram images. It also covers a brief description about the characteristics of the pathologies in the breast tissue and the associated features to help identifying the type of abnormality. A general review of the commonly used algorithms on application of FD for classification of breast lumps as benign and malignant will be provided at the end.

9.1 Introduction

Breast cancer is the second most common type of cancer and one of the major causes for non-accidental death in women worldwide. According to the International Agency for Research on Cancer (IARC), the specialized cancer agency of the World Health Organization (WHO), nearly 1.7 million women were diagnosed with breast cancer in 2012 [1]. This number accounted for 11.9% of the total number of cancer cases occurred in that year, showing more than 20% increase since 2008. Therefore, this growing pattern raises the alarm for breast cancer as a global burden as well as the need for robust screening and control measures to be put in place.

In order to evaluate breast diseases, a number of imaging tests are commonly performed. The tests include mammography exams, ultrasounds and Magnetic Resonance Imaging (MRI) of the breasts, among which the mammography exam (also called mammogram) is the gold standard and one of the most effective tools for breast cancer screening and early detection.

Mammograms are routinely used for early screening of breast cancer and other pathologies in women experiencing no specific symptoms before they can be found by physical examination. It can also be useful for identifying the type of symptomatic breast diseases such as the lumps, cysts, and fibroadenoma as well as differentiation between benign and malignant changes to breast tissue. In this technique, images of the breast tissues are obtained using low-energy x-ray at very safe intensity level. The images produced will be then reviewed and interpreted visually by a radiologist for identification of any suspicious lumps and abnormalities. However, often visual examination of mammograms can be affected by the examiner's error, leading to high cost and low reproducibility of the procedure. Therefore, alongside visual examination, recent advances in digital mammography and image processing techniques have allowed for automatic detection of abnormalities in breast tissue which facilitate early detection of breast cancers and play an important role in reducing the associated morbidity and mortality rate.

Computer based diagnostic systems are being developed to overcome the above shortcomings and make the procedure widely accessible. Such systems also enable quantitative characterization of breast lumps in terms of the size, shape and gray-scale complexity which would not be possible through qualitative visual examination. This is done by extraction of a set of features based on a pattern recognition scheme and analysis of the contours of breast masses. Based on the type and regression level, these lumps show a certain degree of complexity and irregularity in their contour, vascular architecture and texture. Therefore features such as FD can potentially be used to characterize the breast lumps for classification between benign masses and malignant tumors.

9.2 Mammography and Properties of Breast Tissue

The breast refers to the tissue on front side of the chest muscles and is largely composed of fatty tissue. In female body this organ is made up of the milk producing (mammary) gland to produce milk for baby feeding. Milk production does in fact occur in lobules, a sack-like structure within the mammary gland often called "glandular" tissue. The lobules are connected to a complex network of branching ducts (tubes) through which the produced milk is transferred out to the nipple. All these tissues are held in place with the help of the fat and another tissue called the "fibrous". Depending on the density of glandular, fibrous and fatty tissues, the breast density can vary from one person to another. Glandular and fibrous tissues add up to the density of the breast while the fatty issue behaves in an opposite way. The denser the tissue of the breast is, the harder it would be to detect tumors through mammography technique. Therefore in mammography technique, the breasts are slightly compressed between two plates to make the tissue layers thinner and to keep them still. The thinned breast layers will allow for reduction in the x-ray exposure without affecting the quality of images and help increase the sensitivity of detecting tiny calcifications that can be a sign of cancer.

In Mammography, in order to capture an image, the x-ray is passed through the breast tissues form one side (i.e., Top) and gets collected at the other side (i.e., Bottom) while it is being compressed between the two paddles. The collected radiations get recorded on a radiology film or a digital image sensor depending on the model of the machine. The x-rays are absorbed differently by different types of tissues resulting in variations in the intensity of the collected radiations. In the image, they appear as varying shades of white and gray pixels which correspond to the details of the underlying tissues. In general, dense tissues such as bones, calcium deposits, tumors and abnormalities absorb much of radiations and appear white while softer tissues such as muscles, fat and organs allow more rays to pass through and show up in varying shades of gray. Figure 9.1 compares two different mammogram images, one with no abnormality (Fig. 9.1.a) and the other showing a malignant mass (Fig. 9.1.b) with distinctive feature compared to the surrounding normal tissues.

There are a number of known characteristics for the abnormalities in the breast tissue which help the radiologists to interpret mammograms and identify benign from malignant tissues. Most of the changes appear as local variations in density of a region in breast and could be due to normal hormonal changes which are not of any major concern. However,

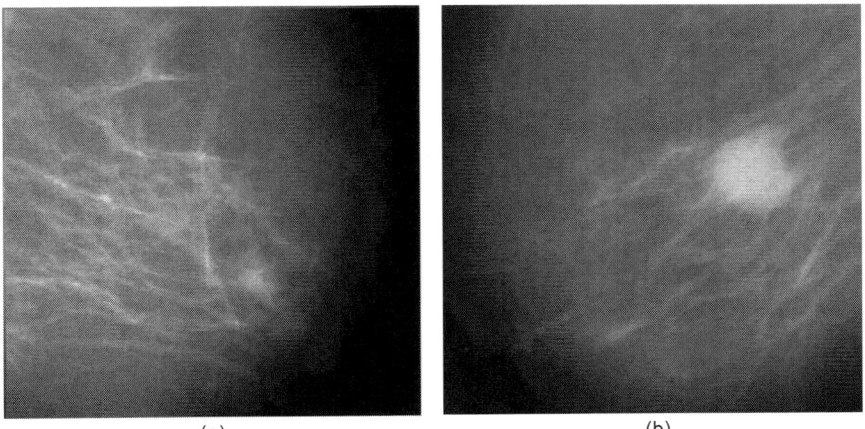

(a) (b)

Figure 9.1. Comparison between two different types of tissues in mammogram images. (a) Normal tissue (b) Malignant mass.

characteristics of certain suspicious masses need to be examined carefully by radiologists before any decision is made.

In a normal mammogram, breast tissue will appear as varying shades of gray intensities corresponding to the ducts, lobular elements, fibrous structure and fatty tissue each having different radiographic densities. These variations may also be different between the two breasts of a single person without being indicator of any abnormality or risk for future complications. In case of any abnormality, they appear as a mass or cluster of calcifications or combination of both. These are mostly characterized on the basis of their texture, location where they appear and whether the patient experiences pain in their breast. The followings are a number of rules, a combination of which will help identification of benign from malignant breast conditions.

- Masses with irregular and random shapes are more likely to be cancerous compared to the ones with smooth round or oval shapes.
- A mass with ill-defined and blurry borders in mammograms is more suspicious for malignancy however, most (not all) benign legions and tumors have a sharp and well defined margin.
- The relative density of a lump to the rest of breast tissue can also be a sign of malignancy. High density tissue and low evidence of fat within a lump increases the risk for presence of cancerous cells, compared to the condition when the density of the tissue is low.
- Malignant breast masses are usually firm, single and hard to move within the breast tissue however, the benign lumps are tender and feel like they can easily move.
- The physical location where the breast lumps occur is also an important diagnostic factor. Cancerous lumps in breast mostly appear either in ducts or in the lobules known as ductal and lobular carcinomas respectively. The names come from the location in the breast where the both types originate. However, they are not considered as invasive cancer unless they have spread out of their place into the surrounding tissue. Therefore, as long as they stay *in situ* (in place), there will be no risk of life-threatening complications for the patient. If the cancer spreads out from its original site and metastases to other tissues such as lymph nodes or other parts of the body, it is said to be in the invasive stage. At this stage the cancer tends to have good prognostic characteristics. For instance invasive ductal carcinomas usually have hard texture with irregular star-like shape and varying cell features [2].
- Pain may also be used as a sign for characterization of breast lumps. In general, cancerous lumps are not painful to touch but in contrast, a painful lump is often indicator of a non-cancerous growth and benign tumor.

Understanding the above characteristics not only helps radiologists determine the different types of lumps but is also essential for engineers to develop computer aided diagnostic systems for automatic characterization of mammograms. However, it should be noted that sometimes these rules alone are not sufficient to rule out the malignancy of the lumps. Further tests such as ultrasounds and an examination of tissue biopsy under a microscope is required to conclude whether a lump is cancerous or not [3].

9.3 Fractal Irregularities of Breast Tissues

As stated earlier, different types of breast lumps including benign masses and malignant tumors exhibit different characteristics in their contours, shape feature and textural properties. Therefore FD has been widely used as a feature for characterization of the shape, border irregularities and gray-scale complexity of the lesions mainly for classification of benign and malignant tumors in mammograms. FD measurement is performed on the border of a breast mass either as a 2D entity or as a 1D profile obtained from the 2D contour. Therefore identification of the contour of breast lesions is the first step towards the calculation of FD. The accuracy of the analyses highly depends on the accuracy of FD measurement which itself results from the precision of contour detection. It also depends on how much detail is obtained through image segmentation and boundary delineation process. However, the presence of noise and large intensity variations in the background of mammogram images makes the segmentation a challenging process and not robust enough for screening and diagnostic applications.

Several approaches have been reported so far for mammogram segmentation and breast lump boundary detection some of which have been briefly described in here:

The earlier techniques were based on simple edge detection algorithms. However, although breast lesions in most digitized mammograms are fairly obvious by visual examination, simple edge detection will not be always effective due to low contrast at the edge points and high portion of fibrograndular tissue to fatty tissue in the background (in some cases). Therefore, Petrick et al. [4] proposed density-weighted contrast enhancement (DWCE) filtering prior to edge detection for enhancement of low contrast objects and to suppress intensity variations in the background. This allowed for a more effective application of a simple edge detector such as the Laplacian-Gaussian for identification of object boundaries.

Zhang et al. [5] suggested application of a modified Region Growing (RG) algorithm to segment mammograms. Unlike the conventional RG techniques in which a constant threshold is used in the contour filters, they proposed adaptive thresholding of gray level intensities for improved controlling of the procedure of adding candidate pixels to the actual boundary pixels. However, a drawback of RG methods is that it relies on the feature homogeneity criterion which requires at least one feature in the ROI to stay uniform for all the pixels [6].

Catarious et al. [7] developed a linear segmentation routine to classify gray-scale pixels of mammograms into interior and exterior classes belonging to breast lumps and background region respectively. The classification was performed using Fisher's Linear Discriminant Analysis (LDA) to determine the optimal threshold between the two pixel classes and outline the objects' boundary.

More recent and robust techniques for mammogram segmentation are based on active contour models [8,9] also known as the "snake". The snake is energy-minimizing spline which is pulled towards the object boundary by external image forces to lock into the edges [10]. In general, active contour models require an initial boundary of the object to be found using a simple edge detection technique as discussed above. This approximate boundary is then used as an initial set of points (also known as seed points) in the active contour model to adaptively lock into the object boundary. It uses prior knowledge about probability of the edges by statistical modeling of the edge features to adaptively tune the parameters of energy minimizing functions. Therefore, snakes have been effective for segmentation of suspicious lesions in mammograms especially when applied to the noisy images with weak lesion boundaries.

For more details and a comprehensive review on the existing techniques for the automatic segmentation of mammogram images, the authors of this book would encourage the readers to refer to work by Oliver et al. [6].

After the edges are determined, FD can be applied to the contour to characterize the degree of complexity of a contour and its association with malignancy of the lump.

As an example, Fig. 9.2 shows different types of breast lumps (after image segmentation) according to their shape and margin. As it can be observed from the image, in general, the contours corresponding to malignant tumors have larger irregularities, variations and shape randomness compared to benign masses. The corresponding 2D box-counting FD values which have been calculated for each contour will also confirm this observation. The FD was obtained after applying morphological skeletonization to the boundary image. Skeletonization was applied to the inverted binary images in which the contours were shown in white on a black background.

For a better comparison, the variations in FD of the above contours have been presented graphically using a box and whisker plot (Fig. 9.3). According to this graph, on average, FDs corresponding to the contour of benign lumps have lower values compared to that of malignant tumors. This confirms the presence of more irregularities and randomness at the boundaries of malignant tumors and that FD can be used to quantify such differences. The result from this example is also in agreement with

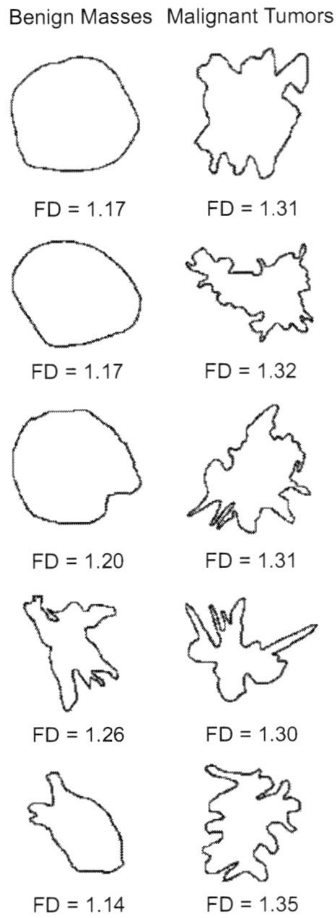

Figure 9.2. Comparison between two different type of breast lumps (i.e., benign masses and malignant tumours) in terms of the shape and border irregularities and the corresponding box-count FD value.

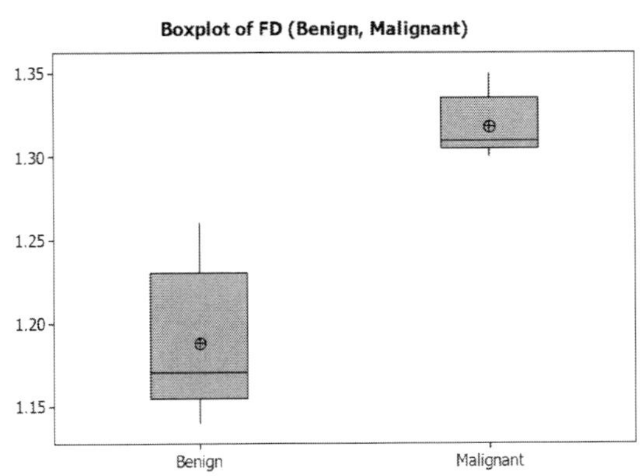

Figure 9.3. Graphical comparison between the variations in FD of two different type of breast lumps (i.e., benign masses and malignant tumours).

the work by Raguso et al. [11] in which the authors have reported a statistically significant difference between the shape complexity of benign and malignant tumors as well as the possibility for classification of breast masses by using FD of their contours.

9.4 Fractal Based Detection of Breast Cancer and the Tumor Types

Automatic analysis of mammograms for the classification of breast cancer and tumor types involves the following four sequential tasks.

1) Detection of the location of suspicious lesions
2) Image segmentation and extraction of the contour of suspicious lesions
3) Identification of the features that best characterize different types of lesions
4) Classification of the regions as benign and malignant using the extracted features

On the basis of observable shape and textural differences between benign and malignant lesions as discussed in the previous section, studies have proposed several features to characterize such differences. Multiple features and descriptors such as compactness, fractional concavity, Fourier descriptor factor have been proposed to be used along with FD for better detection of Breast cancers.

The compactness feature is defined as measure of the efficiency of a contour to enclose a given area [12] which can be quantified using (9.1)

$$C = 1 - \frac{4\pi S}{D^2} \tag{9.1}$$

where D and S are the perimeter and area of the contour respectively.

This feature gives an indication of how round the shape of a contour is. Low compactness represents a more round or oval-like shape-like corresponding to shape of benign masses. Also, considering a contour has a number of concave or convex segments; another parameter which describes the shape of tumors is "fractional concavity". It is defined as the ratio of the cumulative length of the concave segments to the total length of the contour [12]. Large fractional concavity can be related to presence of malignant tumors with a longer cumulative length of the concave segments compared to benign masses.

Fourier descriptor is another useful feature that measures the roughness or high-frequency components in the contours [13]. It describes the shape in 2D space in terms of its spatial frequency using the Fourier method. The advantage of this feature is that it is invariant to rotation, scale and translation effect. The presence of high frequency content in the spectra of the contour will represent rapid changes in the *x* or *y* coordinate at some locations along the contour which can be associated with degree of randomness and irregularities in the shape of malignant tumors. To obtain this feature, each contour is described as two parametric equations as in (9.2) and (9.3):

$$x(i) = x_i \tag{9.2}$$

$$y(i) = y_i \tag{9.3}$$

where (x_i, y_i) refer to the Cartesian coordinate of ith pixel along the contour. Suppose the length of the contour of a breast lump is *N* pixels labeled from *0* to *N-1*. As there is a finite number of discrete contour pixels, taking Discrete Fourier Transform (DFT) of the equations (9.2) and (9.3) will result in two frequency spectra called the "Fourier descriptors" as in (9.4) and (9.5).

$$a_x(f) = DFT(x(i)) \tag{9.4}$$

$$a_y(f) = DFT(y(i)) \tag{9.5}$$

Alternatively, these equations can be re-written by considering the (x_i, y_i) coordinate of ith pixel in the complex plane as in (9.6).

$$z(i) = x(i) + jy(i) \tag{9.6}$$

where *z*(*i*) represents the sequence of contour pixel coordinates in the complex plane with *i* being the number of pixels (i.e., $i = 0,1,\ldots,n-1$). The Fourier descriptor and its normalized version can then be defined as in (9.7) and (9.8) respectively:

$$Z(k) = \frac{1}{N}\sum_{n=0}^{n-1} z(i)\exp[-j\frac{2\pi}{N}ik] \tag{9.7}$$

$$Z_0(k) = \begin{cases} 0 & k = 0 \\ \dfrac{Z(k)}{Z(1)} & k \neq 0 \end{cases} \tag{9.8}$$

where *k* varies within the range of $[-\frac{N}{2}+1, \ \frac{N}{2}]$.

The Fourier descriptor factor, also being referred to by shorter terms as the "Fourier Factor" (FF) [14] can be obtained using equation (9.9) which results into a value in the range of [0, 1].

$$FF = 1 - \frac{\sum_{k=-\frac{N}{2}+1}^{\frac{N}{2}} \frac{|Z_0(k)|}{|k|}}{\sum_{k=-\frac{N}{2}+1}^{\frac{N}{2}} |Z_0(k)|} \tag{9.9}$$

In most studies the above features and several other descriptors (i.e., shape, texture, etc.) have been used simultaneously to obtain a higher classification accuracy compared to the time when only a single feature such as the FD is used. Therefore, limited studies have been conducted on classification of breast masses based on fractal analysis alone as the only shape descriptor. Sole application of FD will reduce the classification accuracy and performance. However, the rest of this chapter will focus only on the application of FD and the literatures on fractal analysis of breast lumps. Other shape descriptors such as compactness, fractional concavity, Fourier descriptor are beyond the scope of this book.

Using the fractal concept, Matsubara et al. [15] proposed an automatic scheme for detection and analysis of breast masses in mammograms. At first, image segmentation was performed to extract breast region by looking into the changes in the density profile. Then the images were categorized into four classes of (i) grandular and fatty (ii) fatty (iii) dense and (iv) high density using analysis of the image histogram and the pectoralis muscles were removed from the region of interest (ROI). The candidate masses were detected by using the intensity thresholding technique while setting different levels for each of the above four categories to extract the boundaries. The candidate regions with a size smaller than a certain value (i.e., 5.6 mm) were discarded from the analysis. Differentiation between types of breast lumps were done by obtaining a series of FD values each corresponding to the contour a candidate mass and monitoring the change in the FD values. This technique resulted into classification accuracy of 100% among seven benign and six malignant cases showing the effectiveness of FD for this particular application. However, no information was provided on the FD range and how it varies with respect to different types of breast lump.

Pohlman et al. [16] used FD together with another five different morphological shape descriptors to distinguish between benign and malignant breast tissue. 51 mammogram images with known types of pathology were segmented based on the RG scheme and the lesions were isolated from the surrounding tissue. They obtained a classification accuracy of around 80% for the FD-based analysis of the contours and found all of the six shape descriptors (including FD) as useful features for diagnosis of the malignancy of breast lumps.

Beheshti et al. [17] proposed a classification technique based on a novel asymmetric fractal features. Mammogram images were first enhanced and the edges of the lumps were sharpened using wavelet decomposition. This was done by the extraction of high frequency components from the wavelet transform followed by application of a piecewise linear gain adjustment to amplify edge of the lumps. A number of new fractal features were extracted which provided a good explanation for the roughness in mass contours as well as the extent of spiculation or smoothness of the masses [18]. FD was also used for detection of breast abnormalities by determining a threshold level that best differentiated masses from the background region. Classification of the lesions as benign and malignant was done using a support vector machine (SVM) classifier with the fractal features given as the training set.

Zhen et al. [19] proposed FD analysis in conjunction n with an artificial intelligent (AI) algorithm for tumor detection in screening mammograms. To compute FD, as the measure of roughness, they divided a given mammogram into blocks of size 16 × 16 and used that measurement to identify the regions containing a tumor. In this work calculation of FD was done using the Blanket method [20] in which FD of fractal curves/surfaces are computed using the area/volume at different scales [21]. Following the same notations as used in [19], the area of the surface (A) at scale r was defined as:

$$A_r = K \times r^{2-D} \tag{9.10}$$

where K is a constant and D is the FD of the surface corresponding to the slope of A_r plot vs. r on a log-log scale as in (9.11).

$$\log(A_r) = (2 - D)\log(r) + k' \tag{9.11}$$

For mammogram lesions as surface area, parameter D is between 2 and 3 however this parameter will range between 1 and 2 for fractal curves. The rougher the curve/surface is the higher the value of FD will be.

In this work, FD was computed to characterize the texture and as a preprocessor to mark the coordinates of the pixels belonging to regions suspicious for cancer and was limited between 2.6 and 2.7.

In order to classify the lesions, mammogram images were then segmented to generate a set of features by introducing the discrete wavelet transform (DWT) based multi-resolution Markov Random Field (MRF) algorithm. Multiple wavelet decomposition was performed to obtain images at different resolutions followed by taking the approximation detail (LL) sub-band for MRF segmentation. Using FD feature, the LL sub-band corresponding to lowest resolution which did not contain any tumor was discarded. Classification was performed based on six different criteria including "edge distance variation", "mean gradient of region boundaries", "mean intensity difference", "area size", "compactness", and "intensity variation". These six features were then given to a binary classification scheme and decision tree to classify the regions as suspicious and unsuspicious. This technique was tested on 322 expert annotated mammograms with classification sensitivity of 97.30% and low false positive and negative rates of 3.92 and 0.0315 respectively.

9.5 Summary

Different type of breast lumps (i.e., Benign masses and malignant tumours) exhibit different characteristics in their contours which can be identified though mammography imaging techniques. The differences are mainly in the shape, border irregularities, gray-scale complexity and density of the lesions. Understanding of these characteristics will help the radiologists to interpret mammograms and identify benign from malignant tissues.

Due to presence of such characteristics, FD can be used to quantify the border irregularities of breast lesions either as a 2D entity or as a 1D profile and has found application in detection of breast cancer. Therefore, this chapter has discussed the application of FD as a useful feature for automatic classification of breast cancer and identification of different types of breast tumours in mammograms. Except FD, other features and shape descriptors such as compactness, fractional concavity, Fourier descriptor factor which can help improve the classification performance have also been introduced.

References

1. Ferlay, J., I. Soerjomataram, M. Ervik, R. Dikshit, S. Eser, C. Mathers, D.M. Parkin, M. Rebelo, D. Forman and F. Bray. 2012. Cancer Incidence and Mortality Worldwide: IARC CancerBase No. 11 [Online]. Available: http://globocan.iarc.fr.
2. Dillon, D., A.J. Guidi and S.J. Schnitt. 2014. Pathology of invasive breast cancer. *In*: Harris, J.R., M.E. Lippman, M. Morrow and O. CK (eds.). Diseases of the Breast. 5th ed. Philadelphia Wolters Kluwer/Lippincott Williams & Wilkins Health.
3. McKenna, Sr., R.J. 1994. The abnormal mammogram radiographic findings, diagnostic options, pathology, and stage of cancer diagnosis. Cancer, 74: 244–55.
4. Petrick, N., C. Heang-Ping, B. Sahiner and W. Datong. 1996. An adaptive density-weighted contrast enhancement filter for mammographic breast mass detection. Medical Imaging, IEEE Transactions on, 15: 59–67.
5. Zhang, H., S. Wei Foo, S.M. Krishnan and T. Choon Hua. 2004. Automated breast masses segmentation in digitized mammograms, in Biomedical Circuits and Systems, 2004 IEEE International Workshop on, pp. S2/2-S1-4.
6. Oliver, A., J. Freixenet, J. Martí, E. Pérez, J. Pont, E.R.E. Denton and R. Ziggelaar. 2010. A review of automatic mass detection and segmentation in mammographic images. Medical Image Analysis, 14: 87–110.
7. Catarious, Jr., D.M., A.H. Baydush and C.E. Floyd, Jr. 2004. Incorporation of an iterative, linear segmentation routine into a mammographic mass CAD system. Med. Phys., 31: 1512–20.

8. Rahmati, P., A. Adler and G. Hamarneh. 2012. Mammography segmentation with maximum likelihood active contours. Medical Image Analysis, 16: 1167–1186.
9. Hao, X., Y. Shen and S.-r. Xia. 2012. Automatic mass segmentation on mammograms combining random walks and active contour. Journal of Zhejiang University Science C, 13: 635–648.
10. Kass, M., A. Witkin and D. Terzopoulos. 1988. Snakes: Active contour models. International Journal of Computer Vision, 1: 321–331.
11. Raguso, G., A. Ancona, L. Chieppa, S. L'Abbate, M.L. Pepe, F. Mangieri, M. DePalo and R.M. Rangayyan. 2010. Application of fractal analysis to mammography. Conf. Proc. IEEE Eng. Med. Biol. Soc., 2010, pp. 3182–5.
12. Rangayyan, R.M., N.R. Mudigonda and J.E. Desautels. 2000. Boundary modelling and shape analysis methods for classification of mammographic masses. Med. Biol. Eng. Comput., 38: 487–96.
13. Rangayyan, R.M., N.M. El-Faramawy, J.E. Desautels and O.A. Alim. 1997. Measures of acutance and shape for classification of breast tumors. IEEE Trans. Med. Imaging, 16: 799–810.
14. Shen, L., R.M. Rangayyan and J.E.L. Desautels. 1993. Detection and Classification of Mammographic Calcifications. International Journal of Pattern Recognition and Artificial Intelligence, 07: 1403–1416.
15. Matsubara, T., H. Fujita, S. Kasai, M. Goto, Y. Tani, T. Hara and T. Endo. 1997. Development of new schemes for detection and analysis of mammographic masses, in Intelligent Information Systems, 1997. IIS '97. Proceedings, pp. 63–66.
16. Pohlman, S., K.A. Powell, N.A. Obuchowski, W.A. Chilcote and S. Grundfest-Broniatowski. 1996. Quantitative classification of breast tumors in digitized mammograms. Medical Physics, 23: 1337–1345.
17. Beheshti, S.M.A., H. Ahmadi Noubari, E. Fatemizadeh and M. Khalili. 2016. Classification of abnormalities in mammograms by new asymmetric fractal features. Biocybernetics and Biomedical Engineering, 36: 56–65.
18. Beheshti, S.M., H. Ahmadi Noubari, E. Fatemizadeh and M. Khalili. 2014. An efficient fractal method for detection and diagnosis of breast masses in mammograms. J. Digit Imaging, 27: 661–9.
19. Zhen, L. and A.K. Chan. 2001. An artificial intelligent algorithm for tumor detection in screening mammogram. Medical Imaging, IEEE Transactions on, 20: 559–567.
20. Mandelbrot, B.B. 1983. The Fractal Geometry of Nature, Henry Holt and Company.
21. Peleg, S., J. Naor, R. Hartley and D. Avnir. 1984. Multiple resolution texture analysis and classification. Pattern Analysis and Machine Intelligence, IEEE Transactions on, PAMI-6, pp. 518–523.

CHAPTER 10

Fractal Dimension of Skin Lesions

ABSTRACT

Skin acts as a protective barrier for the body to keep it safe from the sun's UV rays, infection, harmful substances and loosing too much water. Monitoring the condition of skin is essential for our health and wellbeing. Skin can suffer a number of diseases and traumas, and while many of these are superficial and self-healing, some of these can become life-threatening and debilitating if not treated in the early stages. Timely detecting of these diseases are essential for effective treatment. However, this is dependent on the experience and availability of clinical experts.

A number of techniques are being investigated for determining quantifiable biomarkers that could be used for improving the detection of skin lesions, which may lead to an automatic and assistive diagnosis. Researchers have determined that there is a change in the fractal properties of the skin under different disease conditions and as a result of ageing.

10.1 Introduction

When people refer to things being 'skin deep', the intent is to convey the triviality and in common usage, skin treatment is referred to the beauty industry. However, skin is a very important and complex organ, which serves the purpose of covering the body. Besides the obvious changes that take place with ageing, and weather conditions, it is also impacted by a number of other factors. Health of the skin is essential for the health of an individual. For this purpose, a number of modalities that can describe changes to the skin conditions have been developed.

Skin analyses are performed for a range of applications. These would range from wound management, treatment of burns, artificial skin for prosthesis, cosmetics, detecting cancer and for detecting vasculature disorders. Each of these applications is essential for the health and wellbeing of people. While over the past 50 years, the growth in the modalities for skin analysis has expanded significantly, it is one area of medicine where there is large potential for growth.

Skin can be considered as the largest organ of the body, and definitely the most exposed organ to outside elements. It is exposed to different weather conditions, infections, and traumas and is responsible for being the barrier between the elements and the rest of the body. In the past, there has been a development of cosmetics and other similar products and treatments that provide a temporary change to the skin texture. However, there are a number of skin disorders which need to be diagnosed and treated. Some of the skin conditions that require clinical treatment are; sores, warts, irregular hair growth, hair loss, blisters, chaffing, corns, calluses, and burns.

The exposure of the skin to the elements, insects, allergies, viruses, infections and injuries can lead to conditions such as dead tissue or gangrene, skin inflammation or rashes, and dermatitis or scaly skin, as well as cysts, hives, and other similar conditions. There are also conditions that occur with the presence of parasites such as Scabies, ulcers due to conditions such as diabetes, and varicose veins. While many of the disorders or conditions are self-managed by the body, some of these can become debilitating or even life threatening.

The first step in the treatment process is to diagnose and detect the skin lesion. Visually examining the skin is the most common method of observation, whereas clinicians would also observe the skin condition in terms of texture, and changes to pigmentation. There are then a range of *in-vitro* biochemical and spectral based pathological tests, and *in-vivo* imaging tests such as ultrasound and magnetic resonance imaging (MRI). Recent studies have also determined the changes in the electromagnetic properties of the skin with disease and the relationship of electromagnetic conductivity with the depth of the skin.

Earlier studies and diagnostic techniques have focused on the use of precision measurement tools to obtain the parameters for detecting the lesion. However, such an approach requires high level of manual intervention. Therefore, there is the need for global parameters to identify the regions of the skin which require careful assessment. One such parameter is the chaotic properties of the skin surface, measured based on the fractal concept. This chapter discusses some of the studies which have observed the associations of FD of the skin image with different conditions. For this purpose, the fundamentals of the skin are introduced in the next section.

10.2 Fundamentals of Skin

The skin of each individual appears to be unique. It is an indicator of a number of factors such as gender, age, and race of the person, and is also closely associated with the type of activity undertaken, life style, and even the weather conditions. However, despite these obvious differences, the fundamental properties of the skin are the same for all people.

Skin can be considered as one of the largest organs in human body in terms of mass. In adult men with an average weight of 70 Kg, the human skin weighs 3.86 kg, or 5.5% of the total body mass with a surface area equivalent to 1.7 m^2 [1].

Skin is composed of three main layers namely Epidermis, Dermis and Hypodermis (Subcutaneous fat) layers which have been described as follows.

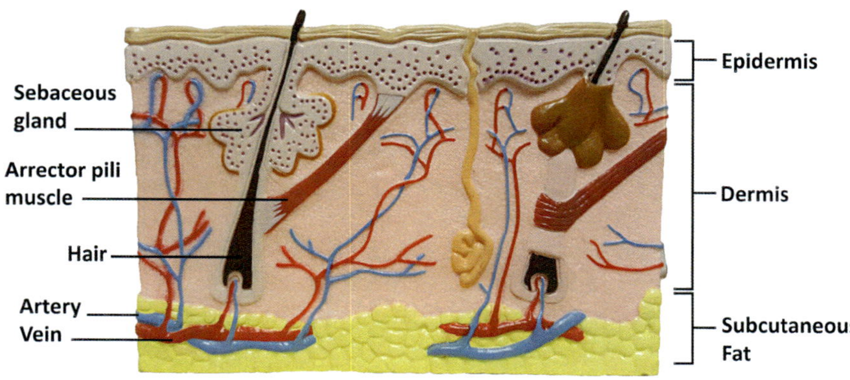

Figure 10.1. Skin structure, different layers of skin.

10.2.1 Epidermis

Epidermis is the outermost layer of the skin. The term "epi" means "over" or "upon" in the Greek language and hence "epidermis" refers to a layer which is over or upon the dermis layer. A healthy and undamaged epidermis forms a protective barrier between the body and the outside world. This includes protection of the inner tissues from bacteria, viruses, toxins and any foreign substance which may come in contact with the skin. It is a relatively thin layer with the thickness varying between 0.05 mm to 1.5 mm depending on the anatomical location. The thickness is higher in the areas such as the sole of the feet which requires stronger protection and better toleration of pressure.

Epidermis contains different cell types including basal cells, squamous cells, and melanocytes each perform a very specific function. Basal cells are small round cells found in the innermost layer of epidermis called the basal layer. These cells constantly divide into new cells and move the older ones up towards the skin surface to replace squamous cells. Squamous cells also called keratinocytes are flat cells found in the outer part of the layer which accounts for around 90% of the cells in Epidermis. These cells produce keratin, a protein which is an important structural element of hair and nails.

At the outer surface, keratinocytes are dead and form a protective barrier for the body to survive. They also play an important role in the healing process of wounds by migrating across the wound bed and filling in the gap created by the wound.

Melanocytes are the cells that produce melanin, the darkening pigment which is responsible for the skin and hair color and is stored in specialized cellular vesicles called melanosomes. Everyone has almost the same amount of melanocytes, however, the size and amount of melanin granules differs immensely in humans with different races and skin colors. Even in a single person the density of melanin can differ from one anatomical location to the other. In some areas such as the palm and soles of the feet where the skin color is lighter, the density of melanocytes and melanin content is much lower compared to areas with darker color.

Melanin also acts as natural sunscreen. It protects the skin's underlying tissues and organs from the damaging effect of ultra violet (UV) radiation of the sun's rays. The photochemical properties of melanocytes allows for production of melanin in response to UV light exposure and prevents subsequent DNA damage and associated life threatening problems such as melanoma (a type of skin cancer). Due to having greater amount of pigmentation and higher melanin content, individuals with darker skin color are at lower risk of developing melanoma than light complexioned people. Therefore tanning, the condition where the skin darkens by production of more amount of melanin, is the natural and protective response of lighter complexion skins to sun exposure and UV light stimulation.

10.2.2 Dermis

Dermis is the second layer of skin from the surface and is located under the epidermis. It serves the purpose of supporting the epidermis or the outer layer. The surface between the dermis and epidermis is rough and includes ridges that increase the surface area between the two layers.

Unlike the epidermis, this layer contains a network of blood vessels. These vessels not only nourish the surrounding skin tissues but also act as a heating/cooling system to regulate the body temperature. With increase in temperature, the blood vessels start to dilate resulting into a larger volume of blood to flow into vessels. As these vessels are close to skin surface, the increase in the blood volume allows for more heat to get released from the skin surface. In contrast, temperature drop will cause the vessels to shrink and restrict the blood flow for lesser amount of heat exchange.

The lower layer of the dermis, the reticular dermis, provides the elasticity to the skin. It has the collagen and elastin proteins which are responsible for the elasticity and flexibility of the skin. The generation of these proteins naturally slows with age and this is one of the causes of skin wrinkle formation. However, there are also other external factors such as radiation, and disease.

The change in the collagen and elastin are observable in the visual texture, and some of the internal details can be obtained with the help of ultrasonography. One important factor in the health of this layer is based on the perfusion of oxygenated blood and measurement of the changes to oxygenated blood as well as the flow of blood in this layer. Change in the oxygen and perfusion can be measured using spectral analysis, and estimated based on the absorption of light at different wavelengths.

Dermis also provides structural and mechanical support and contains nerve endings to provide senses of touch, pain, pressure and temperature; sweat and oil glands to keep the skin soft as well as the hair follicles to produce hair. While the epidermis is the 'dead-skin', the dermis layer has the blood flow. It also has the arrector pili muscle which controls the hair and insulation properties of the skin.

10.2.3 Hypodermis layer

Hypodermis or subcutaneous layer is also known as the fat layer and is located under the dermis. The thickness of this layer varies from person to person and between different anatomical locations. The fat layer acts as a protective cushion against trauma and helps in connecting the dermis to underlying muscles and bones. It also stores energy and provides an insulating barrier to keep body temperature regulated. Like dermis, this layer contains blood vessels and nerve cells with similar functionality. However, in some of the literature, the skin is defined as a two layer organ which covers this fatty layer and the subcutaneous fat is not considered as part of the skin.

This layer shrinks as a result of the normal biological aging process and other conditions. While the precise measurement of this layer is performed using ultrasound imaging, there are number of simple, indirect methods to estimate the thickness of this layer. One of the simplest is using calipers (skinfold test) to measure the fat percentage. While this is only an estimate and not measurement, this process is easy to perform and quite reliable for a large number of applications.

10.3 Skin Lesions and Abnormalities

Skin is often overlooked as a location for a wide range of symptoms, disorders and even life threatening abnormalities. Healthy skin should be well hydrated and have a consistent color, soft smooth texture, uniform appearance and normal sensation with no itching or burning signs. However, sometimes a number of conditions may result in abnormal physiological changes to the appearance of some of the skin compared to the surrounding regions which are known as the skin lesions. Some of these may spread (metastasize) and damage the surrounding organs and pose life-threatening problems. Therefore understanding the clinical features and being able to differentiate between different types of lesions is crucial for the purpose of establishing better diagnostic methods.

10.3.1 Benign abnormalities of skin

In brief skin abnormalities are divided into two general types of benign and malignant legions. The benign lesions account for majority of skin abnormalities and are defined as the skin growths that are neither cancerous nor occur under normal cell division cycle. They usually grow slowly and do not spread to other part of the body. Some examples include:

1) **Freckles:** Small, light to dark brown patches, appearing mostly on the face as a result of long exposure to UV radiations in the sunlight [2].
2) **Moles:** Cluster type growth of melanocyte cells appearing as dark brown spots on the skins. Most moles are benign and do not spread or change overtime however some types are likely to develop into cancer.
3) **Warts:** Dome or flat shaped, cauliflower-like spots, caused by viral infection. Depending on the type they may be pink, yellow, light and even grayish brown with rough surface.
4) **Seborrheic keratosis:** Wart-like growth of spots with appearance similar to melanoma as a type of skin cancer, originates from keratinocytes. They typically vary from yellow to dark-colored wax-like plaque with uneven rough texture.
5) **Hemangiomas:** A mass or lump caused by benign growths of blood vessel at an abnormal rate. It is more common in infants and appears as soft red or blue colored lesions within first few weeks of life.
6) **Psoriasis:** An inflammatory chronic skin condition made up of patches of red, thick and itchy skin that is covered with loose white scales.

10.3.2 Malignant lesions—skin cancer

Malignant lesions are cancerous and often life threatening changes to the skin, they are skin growths caused by disruption in the normal cell division cycle. These lesions are made up of cells that grow and accumulate at a different rate compared to non-cancerous cases. This condition is generally referred to as skin cancer. These cancerous cells can spread to other parts of the body and damage nearby organs.

Skin cancer is classified in terms of the cells that cause the disturbance, or modifications of the cells. Depending on the type of cells that become malignant, the three most common skin cancers are; basal cell, squamous cell and melanoma.

Basal cell skin cancer

Basal cells are located in the lower part of the epidermis. Basal cell skin cancer is the most common type of skin-cancer. The growth of this is slow and is caused by an abnormal growth of Basal cells. It commonly has the appearance as dry and scaly patch. It also may present as firm pearly lump on the region such as head, neck and upper body, which are typically the areas most exposed to UV radiation from the sun. This type of cancer can spread and grow into bones and nerves but it is rarely fatal.

Squamous skin cancer

Basal cells move upwards towards the surface of the epidermis and become flat and become the squamous or scale-like cells. Squamous skin cancer is the second most common type of skin cancer found in epidermis. The symptoms include thick and red plaques, scaly patches and a non-healing sore which has the possibility of bleeding.

Melanoma

Melanoma is caused by the malignant transformation of melanocytes, the cells in the skin that produce melanin. Malignant melanoma (MM) or melanoma manifests itself as a dark lesion and most often with an irregular boundary. While not the common form of skin cancer, it is the most aggressive kind. The degree of irregularity and change in the colour are important diagnostic indicators. An example of a melanoma and its border has been shown in Fig. 10.2.

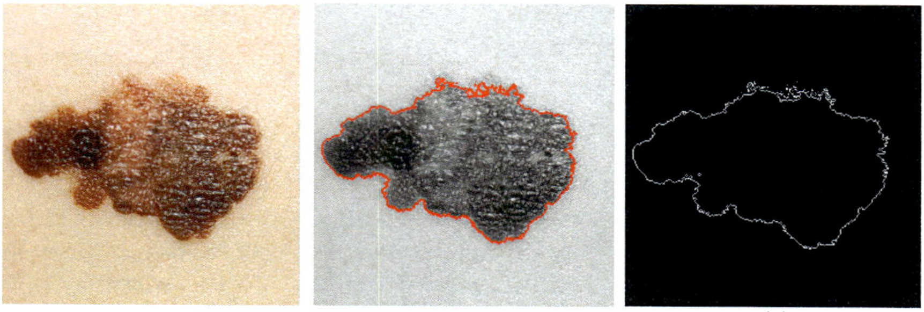

(a) (b) (c)

Figure 10.2. Example of a melanoma and its border irregularities (a) color image (b) gray scale image with highlighted boundary using active contour model (c) boundary of the melanoma in binary format.

Malignancy of a skin lesion and risk for melanoma can be assessed based on a number of visual attributes defined as the ABCD rules of dermatoscopy. This rule which is also referred to as the ABCDE rule allows dermatologist to remember the signs of melanoma and be able to differentiate it from benign lesions. This rule states that the symptoms of melanoma include:

1) A, for a change in the ***Asymmetry*** of the legions so that one half of the lesion is different from the other half
2) B for ***Border*** irregularity such as scalloped, jagged or notched boundaries
3) C for ***Color*** sign including presence multicolored spots
4) D for ***Diameter*** greater than 6 mm
5) E for ***Evolving*** such as change in shape, size and presence of symptoms such as bleeding and itching

10.4 Skin Cancer and Associated Changes to FD

Early identification of skin lesions plays a key role in effective management and treatment of skin disorders. Even malignant legions can be treatable when diagnosed in the early stages, prior to presence of metastasis or secondary complications. Therefore, any noticeable change in the color, texture, distribution, and configuration of the skin is of importance and should not be overlooked. Fortunately most of the skin abnormalities are benign and do not require extensive clinical tests. However, if they are incorrectly diagnosed or remain undiagnosed for a long time they may evolve into further complications.

The very first step for assessment of skin abnormalities is based on the visual examination of the skin by a dermatologist to identify unusual growths and suspicious spots. With advances in image processing and computer aided systems, quantitative characterization of skin legions has become of a great interest. Based on the above observations, a number of research projects have developed methods to automate the assessment of the images of skin abnormalities by investigating the shape of different lesions and their relationship to skin cancer.

There are now commercially available smart phone applications (apps) that automatically process the image to determine if it is mole or malignant melanoma. One company called the "Skin vision" [3] has developed a smart-phone app that performs risk assessment of different types of skin lesions allowing the user to keep track of the change of their skin conditions. It comprises an image processing algorithm based on dermatological evaluation of 4000 cases corresponding to 500 patients with different skin conditions. While such a product is in the early stages, the outcome seems to be promising showing the accuracy of up to 90% for a diagnostic method which is purely based on image processing and examination of the visual attributes of lesions.

As discussed earlier, malignant lesions have significantly rough edges and different textural properties compared with the benign ones. For this reason, the application of FD as a measure of the border irregularities of skin lesions has been tested [4–6] and found to be promising for characterization of different types of skin cancers and classification of them into benign and malignant cases.

Jayalalitha et al. [5] used Box-counting method (FD_{BC}) to measure border irregularities of pigmented skin lesions and the dimension of cancer cells. They reported that this feature is strongly associated with a type of skin abnormality and can be used to determine the cancerous cases.

Carbonetto et al. [6] differentiated between normal moles and melanoma by characterizing their border structure using box-counting FD. They created a feature vector from two points on the FD curve corresponding to FD at scale 2 and 4. The reason for the selection of a limited number of points was to avoid the FD values corresponding to the box sizes which are either comparable to the pixel size or close to the size of entire object on the image. Otherwise it would lead to erroneous approximation of FD and degrade the classification performance.

In this work, classification was performed using Fisher's Linear discriminant analysis (LDA) by projecting the obtained feature vector over a two dimensional space and finding a boundary hyperplane that best divided the data into two classes of normal moles and melanoma cases. The results showed possibility of a good classification performance (i.e., 85%) when using FD and a classifier with a linear decision boundary such as the LDA. The results also indicate a significant difference in the FD of the melanoma compared with a normal mole and a higher level of complexity (i.e., larger FD value) due to increased border irregularity.

However, although FD can serve as an important feature for classification of different types of skin lesions, a more precise diagnosis will require the consideration of other relevant features as indicated by the ABCD rules of dermatoscopy [7].

10.5 FD of the Ageing Skin

Aging is associated with continuous degradation of the functionality of all organs. During the natural aging process, skin is the first organ that shows up visible signs of age-related changes. There exists a number of age associated conditions that commonly affect the skin as people grow older. These can be divided into two groups of healthy (intrinsic) and unhealthy (extrinsic) aging conditions.

In healthy aging, both epidermis and dermis become thinner and more fragile as a result of the reduction in the rate of cell production. This is accompanied by appearance of a number of wrinkles and crepes as well as an increase in paleness and translucency of the skin. The paleness is the result of reduction in the number of melanocyte cells in epidermis. However, the sizes of remaining melanocytes start to increase and form large pigmented areas, also known as aging or liver spots, in some parts of the skin.

Healthy aging also affects the blood vessels in dermis in terms of their functionality, physiology, morphology and visual appearance. As people age, the vessel walls will become more fragile and can easily break. This will lead to some scattered bruising and blood leakage under the skin. It also affects the blood vessel density and vascular organization by impairing the formation of new microvasculature and disorganizing branching geometry of the arterioles and capillaries [8]. This condition can get exacerbated by hypertension and diabetes mellitus as the two most common age associated diseases.

With aging, there will also be a reduction in the number of sweat and oil glands in dermis. Therefore it would be harder for the skin to keep its moisture and cool down. This will result in rough, dry and itchy skin and cause an overheating problem.

Besides the healthy aging signs which are inevitable to happen, the skin may undergo some changes as a result of a disease incidence. These unhealthy changes are related to environmental factors such as sun exposure or diseases that are more likely to happen as people age but are not part of the normal biological aging process. Examples include abnormalities such as benign and malignant (cancerous) skin tumors, scaly skin patches and brown warts as discussed earlier in this chapter.

From the above observations, there are various types of skin changes as a result of aging. These changes will affect the visual appearance, textural properties and complexity of the skin legions. Figure 10.3 provides a comparison between a young and an aged skin where some of the age-related changes and the differences in textural properties can be easily observed. Therefore, fractal analysis can be used as a tool to quantify such alterations.

A number of studies have taken advantage of fractal analysis for human age-group classification based on the aforementioned age-associated changes in skin properties. Yarlagadda et al. [9] studied the possibility of age group classification using Correlation Dimension (CD) of complex facial image, often referred to as a type of FD. They hypothesized that the changes in the CD of the wrinkles as well as the internal bone structure of the face is associated with age factor and therefore can be used for age group classification. In this work the facial RGB image of the subjects were first cropped to cover only the facial region and were converted to gray scale format. This was then followed by application of an edge detection algorithm using the canny operator [10] as optimal edge detector. The

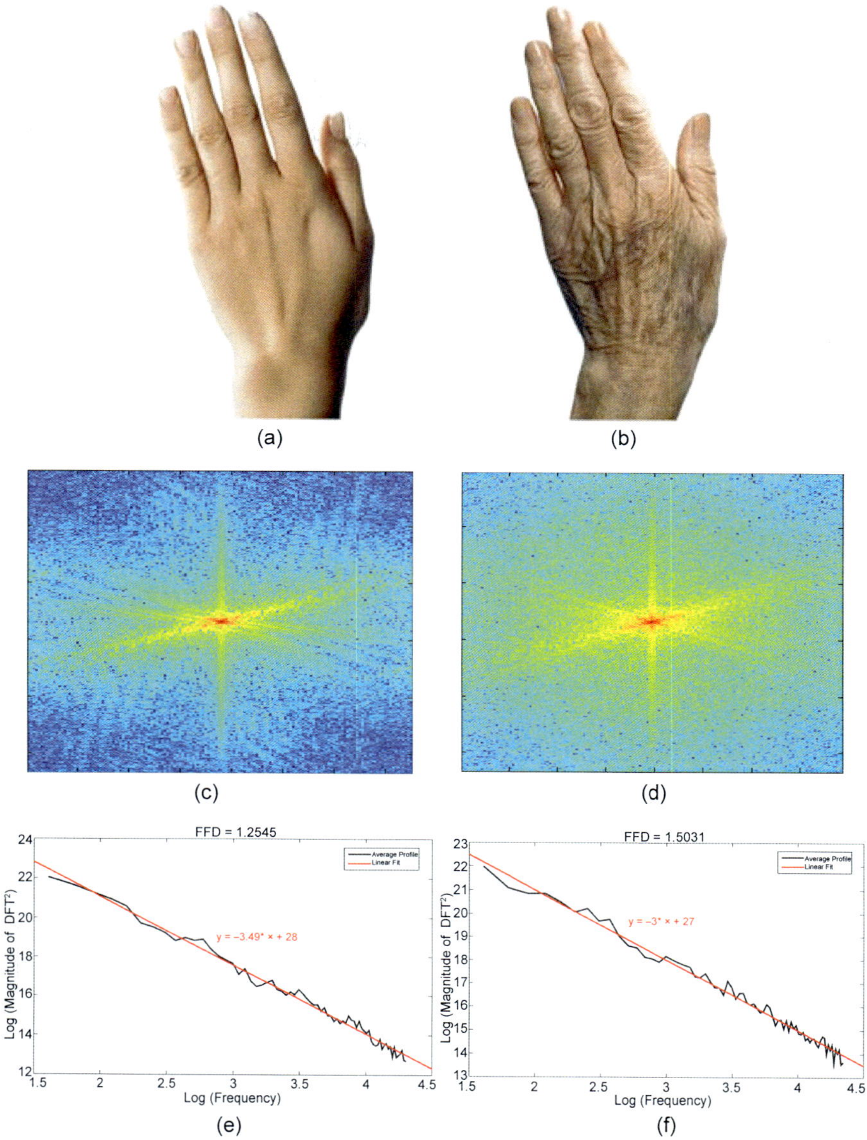

Figure 10.3. Age related FD changes of skin, comparison between (a) young (Left) and (b) old (Right) skin. [Source: http://perfectskinsolutions.co.uk/condition/ageing-hands/] (c) & (d) Magnitude of the power spectrum (e) & (f) plot of average $M(p, q)$ vs. Log (Frequency) and the best fitting line to the data as explained in Chapter 7.

purpose of edge detection was to highlight the significant gray level changes in the image which occur at the boundaries between two regions with large discontinuities in brightness such as the wrinkle edges. The idea was to obtain structure of the facial edges that would change with growing age but remain similar among people with similar age group. This feature was used to calculate the CD of the images and classify them into four distinct age groups of child (0–15), young adult (15–30), middle-aged adults (31–50), and senior adults (>50). Classification was performed based on a simple thresholding scheme considering CD of a child smaller than 1.46, CD of young adult between 1.46 and 1.49, CD of middle age person between 1.49 and 1.54 and CD of a senior adult to be greater than 1.49. Using this technique, they reported a successful classification outcome with accuracy of 99.16%. In this work, the increase in the value of fractal dimension with aging was associated with increased complexity of the skin structure due to appearance of a larger number of wrinkles on the skin.

10.6 Summary

Skin is the covering of our bodies and provides us the protection from the external conditions. It is affected by disease, trauma and ageing and needs to be monitored for the health and wellbeing of the individual. While many of the changes to the skin are only cosmetic, there are also serious and life-threatening changes such as skin-cancer. Most skin ailments diagnostics are performed by dermatologists and other clinician experts, and there is the need for an automated machine based assessment of skin lesions. FD of the skin lesion and quantification of their border irregularities, have been found to be an effective measure of the changes to the skin condition and assessment of the type of lesions for detection of skin cancer.

References

1. Sontheimer, R.D. 2014. Skin is not the largest organ. J. Invest. Dermatol., 134: 581–582.
2. Ogden, E. and J. Schofield. 2013. Benign skin lesions. Medicine, 41: 406–408.
3. Available: https://www.skinvision.com/.
4. Piantanelli, A., P. Maponi, L. Scalise, S. Serresi, A. Cialarini and A. Basso. 2005. Fractal characterisation of boundary irregularity in skin pigmented lesions. Medical and Biological Engineering and Computing, 43: 436–442.
5. Jayalalitha, G. and R. Uthayakumar. 2007. Estimating the skin cancer using fractals, in Conference on Computational Intelligence and Multimedia Applications, 2007. International Conference on, pp. 306–311.
6. Carbonetto, S.H. and S.E. Lew. 2010. Characterization of border structure using fractal dimension in melanomas. Conf. Proc. IEEE Eng. Med. Biol. Soc., 2010, pp. 4088–91.
7. Nachbar, F., W. Stolz, T. Merkle, A.B. Cognetta, T. Vogt, M. Landthaler, P. Bilek, O. Braun-Falco and G. Pelwig. 1994. The ABCD rule of dermatoscopy. High prospective value in the diagnosis of doubtful melanocytic skin lesions. J. Am. Acad. Dermatol., 30: 551–9.

8. Bentov, I. and M.J. Reed. 2015. The effect of aging on the cutaneous microvasculature. Microvasc. Res., 100: 25–31.
9. Yarlagadda, A., J.V.R. Murthy and M.H.M.K. Prasad. 2015. A novel method for human age group classification based on correlation fractal dimension of facial edges. Journal of King Saud University - Computer and Information Sciences, 27: 468–476.
10. Canny, J. 1986. A computational approach to edge detection. Pattern Analysis and Machine Intelligence, IEEE Transactions on, PAMI-8, pp. 679–698.

CHAPTER 11

Case Study I

Age Associated Change of Complexity

ABSTRACT

Earlier chapters have described the fractal properties of different biomedical signals and images. This chapter investigates the comparison between the fractal dimension of the younger and older cohorts. Three case studies have been described; electromyogram (EMG), electrocardiogram (ECG) and eye fundus images. EMG is an indicator of voluntary muscle activity and provides an insight to changes in muscle strength; ECG is for the cardiac activity and is the autonomous activity, while eye fundus is a medical image. The results show age associated reduction in FD for these three recordings. The results also show that the results are sensitive to the method of measuring FD. In conclusion, the case studies confirm that there is a reduction in the FD with age; however it is important to use the appropriate method to measure the change.

11.1 Introduction

Ageing is a natural process during which our bodies undergo significant changes. It is well accepted that age causes weakness in our muscles, it reduces the precision control of our limbs, and this reduces the speed of our actions. Research has also identified that older people are more susceptible to certain diseases in comparison to their younger cohort, and their sensory sensitivity is also reduced. Recent research has now shown a more fundamental change in the body; as we age, the complexity of the physiological or behavioral control system reduces.

It has been proven that when a system is left on its own, there is an increase in entropy with passage of time. However, when the body ages, there is reduction in the complexity of biological systems, which appears to be a contradiction to the fundamental physical laws, for example the second thermodynamics law states that 'whenever energy is transformed from one form to another form, or matter moves freely, entropy increases'. However, careful evaluation shows that there is no contradiction between the two, and this has been elaborated on later.

Many research studies have now revealed that as we age, there is a reduction in the complexities in the brain, heart and the eye [1–5]. Associated with the ageing process, there is also the reduction in the complexity of the cardiac cycle, and our gait pattern. Thus, as we age, there is a net reduction in the number of independent components. In this chapter we report some case studies related to complexity analysis with respect to the physiological changes in the body, particularly in muscle, heart activity and eyes.

11.2 Physiological Basis

The biological systems have been found to be based on simple structures that are repeated over multiple scales leading to seemingly complex and functional systems. These systems demonstrate chaotic behaviour where the outputs can be highly sensitive to the initial conditions, and the system exhibits scale invariance (approximately) properties [1].

In physics, it has been proven that all macro bodies when left isolated, exhibit an increase in entropy. This phenomenon is captured in the second law of thermodynamics. Based on this law, it would be expected that with time, the biological bodies will have an increase in entropy. However, what is observed is the opposite; that as we age, there is a reduction in complexity and entropy. This is however not a contradiction when carefully analysed.

The normal physiological systems exhibit structural and functional complexity. However, ageing and disease are not situations where the biological system is isolated, but it is influenced by the exterior conditions. In such a situation, the laws governing the isolated body are not valid, and the complexity of the structure is governed by the net influences by the exterior conditions such as disease or trauma. In other cases, these conditions may also be the inherent properties of the structure where there is a time related reduction in number or size of cells and thus such a system cannot be considered to be isolated. An ageing system cannot be considered to be isolated except for a very brief period of time.

The quantification of complexity is based on fractal theory. Fractal geometry has been used to study the important aspects of the mechanical function of the various organs [1]. These fractal anatomic structures may show degradation in their structural complexity with aging and disease [1,5]. Aging has shown to be associated with distinctive alterations in the scaling properties and thus a change in the fractal properties. This has been observed significantly in anatomical structures such as the skin, vasculature and the retina of people. It has also been measured in the physiological activities such as brain waves, heart-beat, and muscle activity. It has been reported that aging alters the fractal dynamics of heartbeats and in particular the neurological systems. In this chapter, we discuss some examples and share some of the case studies related to the changes in the muscle, heart and eye due to aging.

11.3 Ageing Muscles and Fractal Properties

Aging has a significant impact on the parameters of muscle functions such as strength, endurance, and fatigue. It is well accepted that muscle strength, speed of contraction and finer control reduce with age, and it is commonly acknowledged that a number of athletic and physical activities such as sprint running are best left to the younger cohort. When performing sub-maximal contractions with hand, arm and leg muscles, elderly adults have a reduced ability to maintain a steady force [6–10]. It has also been seen that age associated weakness of the muscles lead to falls and injuries and there is the need for regular assessment and monitoring of health-parameters for the elderly. However, the underlying principles of the changes are not well understood, with changes attributed to different causes such as loss of muscle mass, changes to muscle fibre types and changes to neural activity. Some studies have investigated the fractal properties of muscle activities and have given new insights to the change.

Recent studies have recommended exercise and lifestyle management to reduce the negative impact of ageing on the bodies so that as we live longer, we should be able to live healthier and more productive lives [1]. For these methods to be effective, it is essential to understand the underlying cause of the changes and design the training and management methods. However, there is a gap of knowledge between number of factors that influence the ageing process and our physical abilities. Age associated changes to human muscle has also been found to be highly variable based on factors such as race and gender. There is also a significant difference between different muscles such as the biceps and Soleus. However, there are some commonalities for example it is well established and well documented that aging is associated with a significant decline in muscle strength and the rate of loss of this strength increases after the age of 60 [6,11].

Aging is considered to have a strong impact on the parameters of muscle functions such as strength, endurance, and fatigue. There is an associated decrease in muscle mass, caused by loss of muscle fibres numbers and decrease in muscle fiber sizes [11,12]. Loss of muscle mass among the aged results in diminished muscle function. Older people have impaired motor performance, slowing of movements, and decrease in muscle strength or maximal force production [10]. This process of aging is highly variable among muscle groups and individuals.

With ageing there are changes in the muscle fibre composition; some of the fast fibres de-differentiate back to slow fibres and there is also reduction in the number of active muscle fibres. This may be the reason for decrease in strength and an increase in endurance in men and women [10]. There are a number of unknowns for example the causes of the changes such as myopathy and neuropathy, and the age at which either of these have a significant effect. It is essential to understand the underlying principle and the study of complexity of the muscle activity is towards better understanding of this phenomena.

The speed of performing actions is slower for the elderly compared to younger people, and older people have reduced speed. Thus their performance of exercises such as sprinting which requires fast actions and large bursts of force are reduced. There are also other changes to the muscle mass and skeletal system leading to reduced flexibility. Psychological changes associated with loss of muscle strength and reduced finer control, have also been observed. However, earlier studies which had not previously been reported on, have measured age associated multi-dimensional changes, such as the relationship of force of contraction, the associated complexity of the muscle activation, and impact of ageing on muscle fatigue. Furthermore, the difference between the ageing processes of muscle control of the two genders has also not been investigated in sufficient detail.

Some of the observations of age associated changes in humans are in terms of change in muscle strength, muscle control, and psychological changes. The grip forces and "safety margins" (grip force in excess of the minimum grip force to prevent slip) of older adults averaged twice that of young adults [8,9] and they demonstrated greater delays between grasping and lifting an object. The functional motor deficits from tactile sensory impairments in old age will vary with task and behavioural context. Previous research studies revealed that either in the seated or in the standing condition, young adults diminished their peak grip force across trials much more than older adults. They also reported that the older adults suffered from somatosensory loss as they reduced their balance instability across trials as much as young adults [6,8,9,13].

Understanding the physiological changes that occur in muscles as a result of aging can be of great importance to not only a physical therapist but to physiologists and other health care providers, and for policy makers. Changes in the neuromuscular system based on the effects of age can be demonstrated by measuring the loss of muscle mass and muscular strength in older adults [14]. Lipsitz and Goldberger [1] proposed that there is a reduction in the complexity of a physiological or behavioural control system with age and disease. They postulated that the reduced complexity reflects the underlying structural (component) and functional (coupling) changes in the organization of the system. A loss in system complexity is reflected by the loss or impairment of functional components and/or due to altered nonlinear coupling between the components [1].

The study by authors [15] has established the association of aging with the change of the fractal dimension of surface electromyogram; sEMG of biceps of the older cohort have a lower fractal dimension compared with the younger ones. The objective of this case study which has been reported in [15] was to experimentally investigate the age associated changes of the muscle activity. For this purpose, surface electromyogram (sEMG) was recorded from normal participants, with the help of non-invasive user friendly recording setup.

11.3.1 Materials

Ninety–six healthy subjects from the Australian urban (multi-racial) population with ages ranging from 20 to 70 years, with no symptoms or history of major neurological or movement disorders volunteered to participate in this study. All the participants were moderately active and performed non-strenuous exercises 3–5 times/week. None of the participants participated in any competitive level sports or regular rigorous exercises. Prior to the start of the experiment, the purpose of the study, procedures and risks associated with participation were explained and written informed consent was obtained from each subject. The experiments were approved by RMIT University Human Research Ethics Committee and were conducted in accordance with Declaration of Helsinki of 1975, as revised in 2004.

sEMG signals were recorded using Delsys (Boston, MA, USA), a proprietary sEMG acquisition system. The system supports bipolar recording and has a fixed gain of 1000, CMRR of 92 dB and bandwidth of 20–450 Hz, with 12 dB/octave roll-off. The sampling rate is fixed at 1000 samples/second, and the resolution is 16 bits/sample. Bipolar electrodes of Delsys (Boston, MA, USA) were placed on the skin of the participant's, overlying the muscle under investigation. These are active electrodes with two silver bars (1 mm wide and 10 mm long) mounted directly on the preamplifier with fixed inter-electrode distance of 10 mm. Two bipolar electrodes were placed on the anterior of the arm above the biceps.

During the experiments, the volunteers were seated on a sturdy and adjustable chair with their feet flat on the floor and their upper arm was rested on the surface of an adjustable desk such that the forearm was vertical. The elbow was maintained at 90 degrees, with the fingers in line with a wall mounted force sensor (S-type force sensor - INTERFACE SM25) attached to a comfortable hand sized ring with a flexible steel wire, and the ring was held on the wrist of the participant. The output of the force sensor was displayed in real time along with an EMG acquisition system.

The maximal voluntary contraction (MVC) was determined by taking the recording three maximal contractions, each 5 seconds in duration and performed with 120 seconds rest-time between each effort. If there were any outliers, the experiment was repeated. After determining MVC, the participants performed isometric contractions at their 75% MVC for the duration of 10 secs. The participants were provided with visual feedback of the force of contraction to assist them in maintaining it constant.

11.3.2 Methods

All the recordings were segmented into 1s segments for the analysis. The initial 3s of the recordings of sEMG and force were removed prior to the data analysis to avoid the initial force stabilization. The segmented data of 5 secs after removal of the initial 3 secs were then analysed to compute Fractal dimension using Higuchi's algorithm explained in Chapter 3, Section 3.4.

11.3.3 Results and discussion

Figure 11.1 shows the change in the complexity (Fractal dimension) of sEMG signal recorded from biceps muscle of the people aged 20 years to 69 years while performing a certain isometric muscle contraction. The results show that the FD reduces with the progress of age. The reduction in FD could indicate a reduction in the number of motor unit, and may indicate the

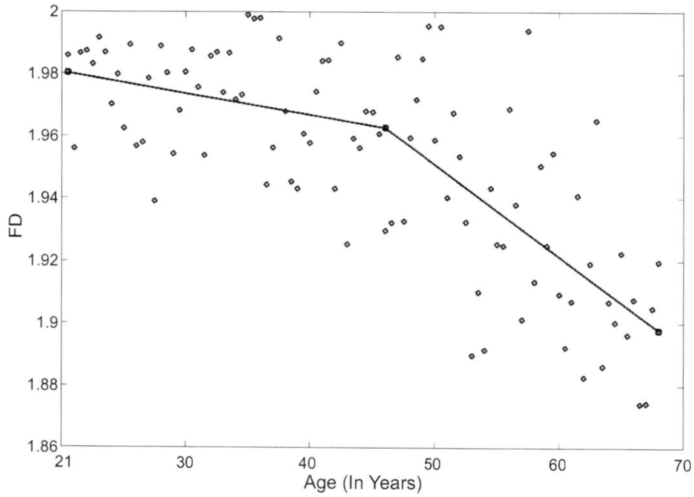

Figure 11.1. Fractal dimension of sEMG recorded from biceps muscle of the people ageing from 20 years to 69 years.

increase in the motor unit density. This is consistent with the findings by other researchers who have studied the change in FD of different aspects of the body with age such as the reduction in FD with the age of the retina vessels and the neurons [2,14,16,17].

11.4 Ageing Heart and Changes to ECG

There are effects of aging that affect the cardiovascular system both structurally and physiologically. Structural changes include increased heart weight, decreased number of myocardial cells with enlargement of remaining cells and decreased vascular tone [18]. Functionally, there is decreased diastolic pressure (during initial filling of the heart); decreased diastolic filling elongation of muscle contraction phase, muscle relaxation phase, and ventricle relaxation [19].

Normal changes in an ECG (or EKG) for an older adult include marginally increased PR, QRS, and Q-T intervals [18–20]. Alongside the changes to the heartbeat, there are also changes to the pressure; systolic pressure may increase due to loss of arterial dispensability, while diastolic remains the same; thus there is an increase in pulse pressure [19]. Although normal pulse rate values change with advancing age, those differences are insignificant. An example to highlight the changes of the cardiovascular system would be; consider a 21-year-old Caucasian male in generally good physical condition and of an average weight and height. The typical expected resting pulse rate would be between 62 and 65 beats per minutes. While this norm fluctuates over the years, the expected heart-rate at age 65 years of an average sized male is still between 62 and 65 beats per minute [18]. However, the resting heart rate could change significantly because of disease and medication effects. Similarly, there can also be significant change to the heart rate with exercise. However, what is important to note is that heart rate continues to change all the time for all of us.

Disease or other conditions can result in abnormal cardiac activity. While in most cases, these abnormalities are for short duration, these may be long term, and are symptoms of disease. It has been found that abnormalities of the heart activities increase with age and it has been reported that approximately 50% of people over the age of 50 year have some ECG abnormalities. The presence of long-range (fractal) correlations in cardiovascular fluctuations in health has implications for understanding and modelling neuro-autonomic regulation. However, this type of scaling behaviour cannot be explained using traditional homeostatic control mechanisms whose goal is to maintain a constant and steady output [21].

Based on traditional medicines from India and China, it is hypothesised that there may be alterations to the scaling behaviour of the cardiac activity. This leads to the question; are pathologic states and/or aging associated with distinctive alterations in these scaling properties? Such information can help explain some of the manual observations made by traditional medicine experts and could be of diagnostic and prognostic use. Investigations have been made by Goldberger et al. [21], where they analysed the scaling behavior of ECG that had been recorded from people suffering from a number of life-threatening cardiac pathologies. The results indicate significant alterations in short and long-range heartbeat correlation properties. This suggests that there may be diagnostic and clinical applications of measuring the scaling properties of the cardiac activities. This can be performed by measuring the fractal properties, or Poincare plots. We have a reported a following case study using the data from a ECG-ID database [22,23].

11.4.1 Materials

ECG data was obtained from The ECG-ID database Physionet [22,23]. The database contains records from volunteers (from 13 to 75 years). For this case study, ECG data from a healthy young (Male, 21 years) and an old subject (Male, 70 years) was analysed. ECG was recorded from each volunteer for 20 seconds, digitized at 500 Hz with 12-bit resolution over a nominal ±10 mV range.

11.4.2 Methods

The recordings were analysed off-line using Matlab platform and the Fractal dimension of the ECG signal was computed using Higuchi's algorithm. The computation of the fractal dimension has been explained in Chapter 3, Section 3.3.

11.4.3 Results and discussion

Figure 11.2 shows the comparison of the fractal dimension of the younger and older subjects. From this figure, it is observed that the fractal dimension of ECG was lower for the older cohort compared with the younger subjects.

11.5 Ageing Eyes and FD of Eye-fundus Images

With age, there are changes to the retina of the eye, and it has been demonstrated that there is an association between aging and FD of retinal images. The association is in reduction in the complexity of retinal vessels as a result of normal aging processes which can be quantified using fractal analysis. The experiments have shown that while the change in FD is observed using spectral based measure of FD, it is not observable using other FD algorithms such as box-counting.

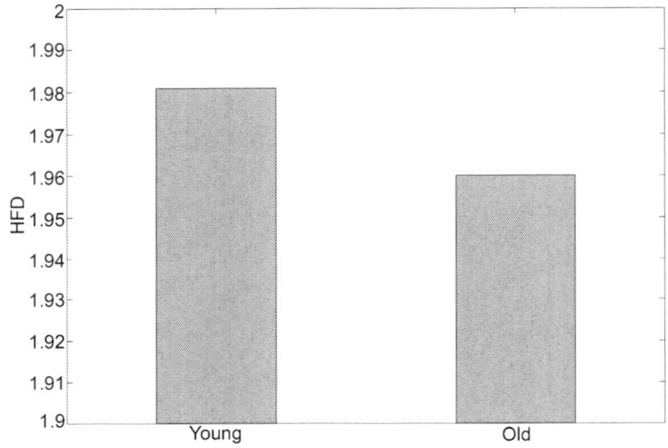

Figure 11.2. Mean Fractal dimension of ECG signal for young and old healthy subjects.

This section provides a case study where the association of FD of the retinal vasculature was measured for younger and older cohorts. The eye fundus images were analysed to obtain the fractal dimension of the retinal vasculature and two FD methods were compared. The methodology and results presented in this section have been obtained from our previous works on this topic [24,25].

11.5.1 Materials

Retinal images were obtained from the Blue Mountains Eye Study (BMES) database. BMES is a population-based longitudinal study of eye diseases and other health outcomes in a suburban Australian population [26]. A Zeiss FF3 fundus camera was used to capture 30 degree field Optic disk cantered retinal photographs. All the images were taken on 35 mm film and in mydriatic mode. The photographs were then digitized using a Canon CanoScan FS2710 slide scanner (Canon Corporation, Tokyo, Japan). From this dataset 400 images of left and right eyes (800 retinal images in total) corresponding to 400 healthy individuals were randomly selected and categorized into four narrow age groups of 50–59, 60–69, 70–79 and 80–89 years, each consisting of 50 men and 50 women. However, 20 right (5%) and 32 left (8%) images were found to be upgradable and therefore were discarded from the analysis.

11.5.2 Methods

The original images of 3888 × 2592 pixels were down sampled to the new size of 778 × 519 pixels to reduce the computation time and make the dimensions comparable with other similar studies [27]. A region of interest (ROI) was manually selected to centre the optic disk region (ODR) and cropped to the new size of 300 × 300 pixels corresponding to 2.5 ODR radii. A wavelet based image enhancement was performed using the previously explained method by Soares et al. [28] to ensure the analyses were not degraded by image artefacts. The enhanced images were represented in gray-scale as a 3D surface for further analysis.

In this work, FFD was computed based in the method explained in Chapter 7 to approximate the FD of retinal images. FDs of the four age groups were statistically compared using one-way analysis of variance (ANOVA). Distribution of FFD data for both left and right eyes was also tested for normality prior to statistical analysis. The data was then dichotomised into the two age groups of 50–69 and 70–89 years to further study the significance of this association.

11.5.3 Results

FFD values were found to be normally distributed with mean ± standard deviation (SD) of 1.511 ± 0.03 and 1.506 ± 0.03 for the left and right eye respectively. For both eyes an overall reduction in FFD was observed with increase in age as shown in Fig. 11.3 across the four age groups. The mean and SD of the FFD values have also been shown in this figure. Linear regression analysis was performed to test for liner trend in FFD as a result of aging by estimating the goodness of fit (R^2). The result showed that reduction in FFD was linearly associated with aging for both left (R^2 = 0.94) and right (R^2 = 0.99) eyes with the average reduction rate of 0.01 per decade of age. This association was found to be statistically significant (P < 0.001) especially when comparison was made between the two age groups of 50–69 and 70–89 with a 20-year interval.

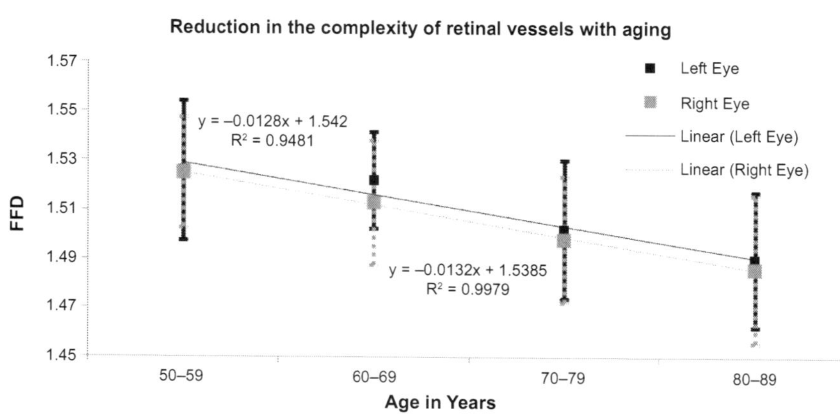

Figure 11.3. Association between FFD of retinal images and aging.

11.5.4 Discussion

In this study the hypothesis of reduction in the complexity of retinal vasculature due to normal aging process was tested and confirmed using FFD. We modelled the change in FFD as an linear function of age and observed average reduction of 0.01 for each decade increase. However, this linear association may not be valid for an older population aged 50 and over. In a similar study we have shown that in a sample with a broader age spectrum (i.e., 10 to 73), reduction in FD is better explained by the quadratic model than by any other models (i.e., linear, cubic) [29]. However, in this study FD_{BC} was used as the measure of complexity of retinal vasculature. These dissimilarities could be attributable to application of FD_{BC} instead of FFD and to the potential error sources as a result of the binarization process.

An important implication of this study is the need for consideration of age as a confounding factor in all statistical analysis and models involving FD parameter, when assessing associations of FD with diseases of interest. This can be done by normalization and adjustment of the data for the age factor.

11.6 Summary

This chapter has described three case studies that compare the fractal dimension (FD) of young and older people; EMG analysis, ECG analysis and Eye Fundus image analysis. The experimental results show that there is a reduction in FD with age for each of these three recordings. The results also show that there are significant differences between different methods for measuring FD. While box-counting is unable to identify any change in the FD of the retinal vasculature, spectral FD captures the difference between the two cohorts. In conclusion, the case studies confirm that there is a reduction in the FD with age; however it is important to use the appropriate method to measure the change.

References

1. Lipsitz, L.A. and A.L. Goldberger. 1992. Loss of 'complexity' and aging: potential applications of fractals and chaos theory to senescence. JAMA, 267: 1806–1809.
2. Azemin, M.Z.C., D.K. Kumar, T.Y. Wong, J.J. Wang, P. Mitchell, R. Kawasaki and H. Wu. 2012. Age-related rarefaction in the fractal dimension of retinal vessel. Neurobiology of Aging, 33(1): 194.e1–194.e4.
3. Brown, M. and E.M. Hasser. 1996. Complexity of age-related change in skeletal muscle. The Journal of Gerontology, J. Gerontol. A Biol. Sci. Med. Sci., 51(2): B117–23.
4. Wong, K.C.L., L. Wang, H. Zhang, H. Liu and P. Shi. 2009. Computational complexity reduction for volumetric cardiac deformation recovery. J. Sign. Process. Syst., 55: 281–296.
5. Kaplan, D.T., M.I. Furman, S.M. Pincus, S.M. Ryan, L.A. Lipsitz and A.L. Goldberger. 1991. Aging and the complexity of cardiovascular dynamics. Biophys. J., 59: 945–949.

6. Bazzucchi, I., M. Marchetti, A. Rosponi, L. Fattorini, V. Castellano, P. Sbriccoli and F. Felic. 2005. Differences in the force/endurance relationship between young and older men. Eur. J. Appl. Physiol., 93: 390–397.
7. Miller, R.G. 1995. The effects of aging upon nerve and muscle function and their importance for neurorehabilitation. Neurorehabil. Neural. Repair., 9(3): 175–181.
8. Cole, K.J. 1991. Grasp force control in older adults. J. Mot. Behav., 23: 251–258.
9. Klein, C.S., C.L. Rice and G.D. Marsh. 2001. Normalized force, activation, and co-activation in the arm muscles of young and old men. J. Appl. Physiol., 91: 1341–1349.
10. Kent-Braun, J.A. and A.V. Ng. 1999. Specific strength and voluntary muscle activation in young and elderly women and men. J. Appl. Physiol., 87: 22–29.
11. Porter, M.M., A.A. Vandervoort and J. Lexel. 1995. Aging of human muscle: structure, function and adaptability. J. Sci. Med. Sport., 5: 129–142.
12. Doherty, T.J., A.A. Vandervoort, A.W. Taylor and W.F. Brown. 1993. Effects of motor unit losses on strength in older men and women. J. Appl. Physiol., 74: 868–874.
13. Galganski, M.E., A.J. Fuglevand and R.M. Enoka. 1993. Reduced control of motor output in a human hand muscle of elderly subjects during submaximal contractions. J. Neurophysiol., 69: 2108–2115.
14. Kadhiresan, V.A., C.A. Hassett and J.A. Faulkner. 1996. Properties of single motor units in medial gastrocnemius muscles of adult and old rats. J. Physiol., 493: 543–552.
15. Arjunan, S.P. and D.K. Kumar. 2013. Age-associated changes in muscle activity during isometric contraction. Muscle Nerve, 47: 545–549.
16. Merletti, R., D. Farina, M. Gazzoni and M.P. Schieroni. 2002. Effect of age on muscle functions investigated with surface electromyography. Muscle & Nerve, 25(1): 65.
17. Monaco, V., A. Ghionzoli and S. Micera. 2010. Age-related modifications of muscle synergies and spinal cord activity during locomotion. J. Neurophysiol., 104: 2092–2102.
18. Larsen, P. and P.R.C. Fngna. 2008. A review of cardiovascular changes in the older adult. ARN Network, 3(9).
19. Ferebee, L. 2006. Cardiovascular function. pp. 468–503. *In*: Meiner, S.E. and A.G. Lueckenotte (eds.). Gerontologic Nursing (3rd ed.). St. Louis: Mosby Elsevier.
20. Jett, K. 2008. Physiological changes with aging. pp. 65–87. *In*: Ebersole, P., P. Hess, T. Touhy, K. Jett and A. Luggen (eds.). Toward Healthy Aging: Human Needs and Nursing Response (7th ed.). St. Louis: Mosby Elsevier.
21. Goldberger, A.L., L.A.N. Amaral, J.M. Hausdorff, P.C. Ivanov, C.-K. Peng and H.E. Stanley. 2002. Fractal dynamics in physiology: Alterations with disease and aging. Proceedings of the National Academy of Sciences of the United States of America, 99(Suppl 1): 2466–2472.
22. Lugovaya, T.S. 2005. Biometric human identification based on electrocardiogram. [Master's thesis] Faculty of Computing Technologies and Informatics, Electrotechnical University "LETI", Saint-Petersburg, Russian Federation.
23. Goldberger, A.L., L.A.N. Amaral, L. Glass, J.M. Hausdorff, P.Ch. Ivanov, R.G. Mark, J.E. Mietus, G.B. Moody, C.-K. Peng and H.E. Stanley. 2000. PhysioBank, PhysioToolkit, and PhysioNet: Components of a New Research Resource for Complex Physiologic Signals. Circulation, 101(23): e215–e220.
24. Azemin, M.Z., D.K. Kumar, T.Y. Wong, R. Kawasaki, P. Mitchell and J.J. Wang. 2011. Robust methodology for fractal analysis of the retinal vasculature. IEEE Trans. Med. Imaging, 30: 243–50.
25. Azemin, M.Z.C., D.K. Kumar, T.Y. Wong, J.J. Wang, P. Mitchell, R. Kawasaki and H. Wu. 2012. Age-related rarefaction in the fractal dimension of retinal vessel. Neurobiology of Aging, 33: 194.e1–194.e4.
26. Mitchell, P., W. Smith, K. Attebok and J.J. Wang. 1995. Prevalence of agerelated maculopathy in Australia. The Blue Mountains Eye Study. Ophthalmology, 102: 1450–60.
27. Gang, L., Chutatape and S.M. Krishan. 2002. Detection and measurement of retinal vessels in fundus images using amplitude modified second-order Gaussian filter. Biomedical Engineering, IEEE Transactions on, 49: 168–172.

28. Soares, J.V.B., R.M. Cesar, H.K. Jelimek and M.J. Cree. 2006. Retinal vessel segmentation using the 2-D Gabor wavelet and supervised classification. IEEE Transactions on Medical Imaging, 25: 1214–1222.
29. Che Azemin, M.Z., F.Ab. Hamid, A. Aminuddin, J.J. Wang and R. Kawasaki. 2013. Age-related rarefaction in retinal vasculature is not linear. Experimental Eye Research, 116: 355–358.

CHAPTER 12

Case Study 2
Health, Well-being and Fractal Properties

ABSTRACT

In this book we have described the fractal properties of biomedical signals and medical images. This chapter gives three examples of the applications of fractal geometry based analysis of biomedical signals and images. These examples describe the association of the fractal dimension with three conditions; diabetes, stroke and fatigue. One significant advantage of using fractal dimension as a feature to associate with disease is that it is a global to the image or the signal, and often can be performed without need for segmentation. This makes it very suitable for automated machine based assessment for diseases and other conditions.

12.1 Introduction

The first 10 Chapters in this book have described the fractal properties of biomedical signals and medical images. This chapter gives three case studies that demonstrate some of the applications for fractal analysis of medical images and signals. While earlier chapters have provided the frame work for the analysis, this chapter describes the experiments and helps in developing a better appreciation of the applications. This chapter shows the association of the fractal dimension of the signal with three conditions; diabetes, stroke and fatigue.

There are a range of features that have been developed to analyse medical images and signals. However, these typically require manual supervision and this makes them unsuitable for automated and machine based analysis. Fractal dimension has the strength that it is a global feature and the analysis can be performed without supervision, thus making it suitable for automated analysis.

In the next section of this chapter, the association of the fractal dimension of the retinal vasculature using eye fundus images with stroke has been investigated. In the following section, the association of fractal dimension of retinal vasculature with diabetes prior to the occurrence of retinopathy has been studied. In the last section, the change in fractal dimension of surface electromyogram with localised muscle fatigue is shown.

12.2 Risk of Stroke and Retinal Fractal

It has been discussed earlier in Chapter 8, Section 8.5, that there is an association between FD of eye fundus images with coronary heart disease (CHD) or an episode of stroke. This section will provide a case study and a more focused analysis in the context of FD and its association to stroke event [1]. The aim of this study was to measure the difference in the fractal dimension of the retinal vasculature between control participants and someone who later suffered from an episode of stroke. The study also compared different methods for computing FD and the influence of this difference on the association.

The study investigated the difference between different fractal measures and three methods were compared; Higuchi's FD, FFD and FD_{BC}. Different ways of scanning the images to obtain the data arrays required to compute Higuchi's FD were also compared. The quality of the retinal images were first enhanced and then analysed with both preserved and masked ODR to study the effect of ODR masking on the extraction of relevant information associated with a stroke event. This work has also looked into the hypothesis that the image artifacts in the optic disk region (ODR) can degrade the analysis, and ODR may need to be masked prior to estimation of the FD.

12.2.1 Materials

The retinal images from the BMES, a population-based study conducted in a suburban region west of Sydney, Australia, were analysed [2,3]. The age range of participant's was 50–89 years. Images of the retina from both eyes of the study participants were obtained using a Zeiss FF3 fundus camera with 30 degree field of view. The photographs were taken after pupil dilation. The images were digitized using a Cannon FS2710 scanner with maximum resolution of 2720 dpi in 24-bit colour format. Among the total number of 1532 optic disk (OD) centred images (3888 × 2592 pixels) in the database, there were 104 confirmed cases with CHD or10-year follow-up after a stroke.

Self-reported stroke events were validated against medical records of physician diagnosis based on the World Health Organization Monitoring Trends and Determinants in Cardiovascular Disease (WHO-MONICA) plus evidence from computed tomography or magnetic resonance imaging [3]. Only images from the left eye were analysed. Eight of the images from the whole database were discarded due to quality and contrast problem. Retinal images of stroke and cases were matched based on the age (mean (SD) = 67.76 (5.72)), metric body mass index (BMI) (26.32 (4.35)), blood pressure (mmHg) (systolic: 150.64 (18.94)), diastolic: 83.40 (10.35)) and history of smoking. An independent third party ophthalmologist confirmed matching of 46 cases with 39 controls which were used in this study. The matching control set was of people who did not have a history of stroke, hypertension, and diabetes at the baseline and did not develop any of them during the period of the study.

12.2.2 Method

The inverted green channel was used for analysis as this provided a better vessel for background contrast compared to the Red and Blue channels [4]. The images were cropped using a mask of size 1960 × 1960 pixels to cover a region of interest (ROI) corresponding to a circle of 4 ODR diameter (D_{OD}) centered at ODR center. The ODR center and boundary were identified manually by the grader [5]. The cropped image was then down-sampled to 400 × 400 pixels to reduce the computational complexity and make the image size comparable to other studies [4,5]. Image enhancement was also performed to compensate for uneven illuminations and ocular media opacity using a 2D Gabor wavelet filter as discussed in Chapter 7, using the online "mlvessel v1.3" Matlab based software provided by Soares et al. [6].

As stated in Chapter 7.7, Higuchi measures the FD of a set of points in the form of 1D time series [7]. It can address some limitations of other 2D fractal algorithms such as the box-counting through obtaining the FD along specific directions (scanning paths). In order to find the FD of the images (as 2D entity), the intensity values need to be transformed into image profiles as 1D signals. This is possible through a number of scanning directions including horizontal and vertical by decomposing an image of size (M × N) into its M rows and N columns as well as the spiral and radial extraction of the intensity values [8]. In this work several scanning methods were tested and the association of each with a stroke event was studied. For the horizontal (Fig. 12.1.a) and vertical (Fig. 12.1.b) scanning, the gray values were scanned along horizontal and vertical lines and a set of FD-$H_{1\text{-}M}$ and FD-$V_{1\text{-}N}$ values corresponding to the number of columns and rows were obtained. The final FD-H and FD-V were obtained by averaging the horizontal (FD-$H_{1\text{-}M}$) and vertical (FD-$V_{1\text{-}N}$) dimensions respectively. Another possible technique for calculating Higuchi's dimension of retinal images was to scan the image along a number of radial lines passing through the ODR center and recording 1D intensities along radial lines equally spaced with θ° angles (here θ was set to 1°) to cover the ROI (Fig. 12.1.c). A set of FD-$R_{1\text{-}k}$ with the *K* being the total number of lines was obtained and the overall FD-R was calculated by averaging the whole values.

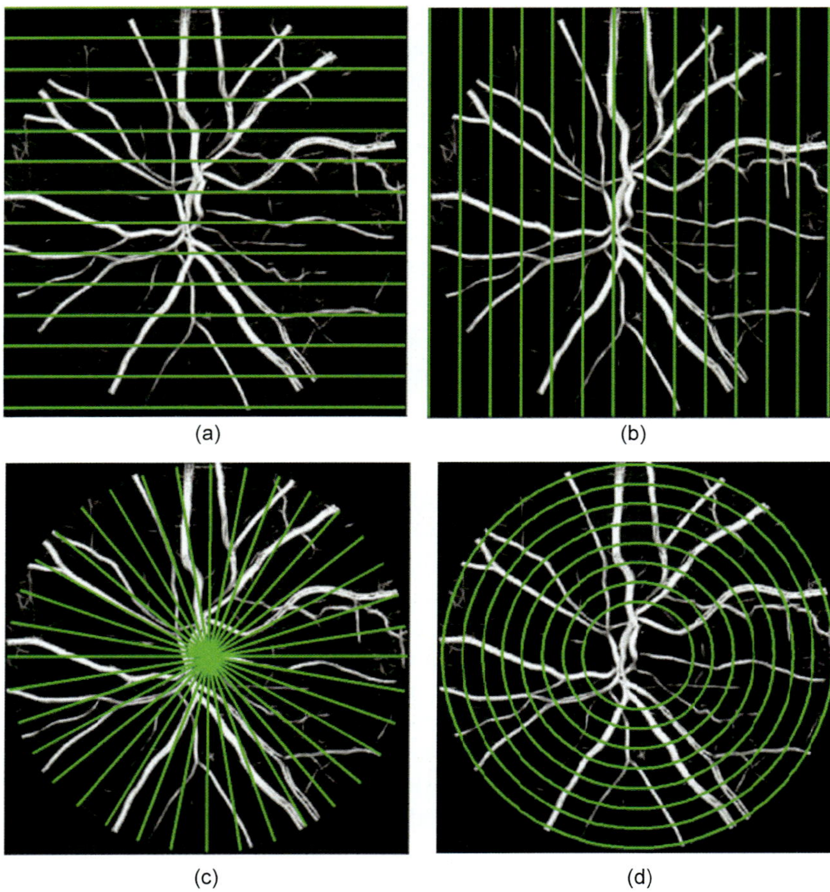

Figure 12.1. Demonstration of different scanning methods (a) Horizontal (b) Vertical (c) Radial (d) Circular on the enhanced retinal image.

Selection of an appropriate scanning method depends on image features of interest. In this study concentric circles around the ODR were proposed and tested. Such a scanning path intersects with most major vessels' along their cross-section and includes vessel caliber information of the major vessels. As such vessels in the retina are in a radial direction with respect to the ODR. The circles were spaced with one pixel intervals in the ROI with the innermost circle having its radius one pixel greater than that of the ODR, and the outermost circle with diameter $4 \times D_{OD}$. Some examples of these scanning circles are shown in Fig. 12.1.d. Higuchi's FD was estimated for each of the circular scans and referred to as FD circular, or FD-C. The average of FD-C of all circles was considered as FD-C corresponding to the ROI.

12.2.3 Data analysis

Statistical analyses were performed using Minitab 16 (Minitab Inc.). For each FD measurement technique, the measured values between the subjects were compared and the significance of association (*p*-values) between case and control groups was tested using the non-parametric Kruskal-Wallis analysis as an alternative method to ANOVA analysis used when the normality condition of the data cannot be satisfied. The Kruskal-Wallis test identifies whether the population medians (η) on response variables are similar across all levels of factors (H_0: $\eta_{Case} = \eta_{Control}$ vs. H_1: $\eta_{Case} \neq \eta_{Control}$) and unlike ANOVA and T-test, the distribution of the response variables do not have to be of any particular form (e.g., normal). This test is suitable for populations with two levels of independent variables [in our case (case/control)].

Two different sets of tests were performed; one on full images and the other on the ODR masked versions to study the effect of ODR on the outcome. The 95% confidence interval (CI) for the difference in population medians ($\eta_{Case}-\eta_{Control}$) was also obtained by non-parametric pair-wise comparison between the case and control groups using the Mann-Whitney test as a follow up measure. The Mann-Whitney test is also known as Wilcoxon rank-sum test and is an alternative to two-sample T-test for measuring the statistical significance between two populations (i.e., case and control) when the data is not normally distributed.

12.2.4 Results

Tables 12.1 and 12.2, show a comparison between different FD measurements and the effect of ODR masking on the statistical significance of the case and control groups. The first column identifies the method by which the FD was calculated. The second and fifth column give the median of FDs for control and case groups. The average rank is presented in the third and sixth columns while Z-values are in columns four and seven. The 95% CI for $\eta_{Case}-\eta_{Control}$ (from Mann-Whitney test) are in column eight, and the Kruskal-Wallis H parameters are listed in column nine. The last column illustrates the *p*-value obtained by Kruskal-Wallis analysis. The results show that there is a statistically significant difference ($H = 5.80$, $p = 0.016$, $\alpha = 0.05$) between FDs of the case and control groups obtained when using the FD-C with 95% CI $\eta_{Case}-\eta_{Control} = -0.0087, -0.0007$. However, no statistical significance was detected for other FDs (all *p*-values > 0.05).

The ODR masking was found to have no effect on the statistical significance between the case and control groups for FD-H ($H = 0.07$, $p = 0.798$), FD-V

Table 12.1. Non-parametric Kruskal-Wallis and Mann-Whitney significance test between control and case groups on retinal images with OD included.

Method	Full Retinal Images (OD included)						95% CI for $\eta_{Case}-\eta_{Control}$	H	**P-Value $\eta_{Case}=\eta_{Control}$ Vs ($\eta_{Case}\neq\eta_{Control}$)
	Control (n = 39)			Case (n = 46)					
	Median (η)	Average Rank	Z	Median (η)	Average Rank	Z			
FD-H	1.851	46.1	1.08	1.845	40.3	–1.08	–0.0093,0.0027	1.16	0.282
FD-V	1.894	45.4	0.81	1.892	41.0	–0.81	–0.0082,0.0034	0.66	0.417
FD-C*	1.987	50.0	2.41	1.981	37.1	–2.41	–0.0087, –0.0007	5.80	0.016
FD-R	1.837	42.9	–0.03	1.843	43.1	0.03	–0.0242,0.0227	0.00	0.979
SFD	1.335	46.1	1.07	1.335	40.4	–1.07	–0.0006,0.0002	1.14	0.286
FD_{BC}	1.648	41.3	–0.58	1.653	44.4	0.58	–0.0080,0.0140	0.33	0.563

Table 12.2. Non-parametric Kruskal-Wallis and Mann-Whitney significance test between control and case groups on retinal images with masked OD.

Method	OD Masked Retinal Images						95 %CI for $\eta_{Case}-\eta_{Control}$	H	**P-Value $\eta_{Case} = \eta_{Control}$ Vs ($\eta_{Case} \neq \eta_{Control}$)
	Control (n = 39)			Case (n = 46)					
	Median (η)	Average Rank	Z	Median (η)	Average Rank	Z			
FD-H	1.841	42.3	–0.26	1.841	43.6	0.26	–0.0067,0.0087	0.07	0.798
FD-V	1.879	47.6	1.57	1.877	39.1	–1.57	–0.0072,0.0008	2.46	0.116
FD-C*	1.987	50.0	2.41	1.981	37.1	–2.41	–0.0087, –0.0007	5.80	0.016
FD-R	1.869	43.2	0.05	1.868	42.9	–0.05	–0.0156,0.0153	0.00	0.958
SFD	1.336	45.1	0.71	1.336	41.2	–0.71	–0.0005,0.0002	0.51	0.475
FD_{BC}	1.623	41.1	–0.64	1.623	44.6	0.64	–0.0078,0.0144	0.41	0.520

* Circular scanning method does not cover the OD area therefore there will be no change on the values of FD-C after masking the OD.

** Bonferronni corrected *P*-values (*N*=6).

(H = 2.46, $p = 0.116$), FD-R (H = 0.0, $p = 0.958$), SFD (H = 0.51, $p = 0.475$) and FD_{BC} (H = 0.41, $p = 0.520$) measurement methods. The overall result for the FD-C methods, shows lower fractal values for the cases compared to the control groups.

12.2.5 Discussion and conclusion

FD is a convenient measure for summarizing the retinal vessel complexity and has found application in case/control studies [9–11] such as for stroke and CHD. Unlike other measurements such as vessel diameter, FD measurement does not require extensive manual supervision and is suitable for automatic assessments and feature summarization. This work compared the application of different FDs to study the association between the changes in FD and stroke event/CHD. To obtain Higuchi's dimension, the image was scanned using different scanning directions among which the circular method (FD-C) was found to be more effective. Unlike other FD methods, FD-C allowed for exclusion of ODR from the analysis, however, in this specific case study ODR masking was found to have no effect on the final result.

The other advantage of this method was that unlike FD_{BC}, it did not require image segmentation [12]. In order to test its efficacy in distinguishing between case and control, it was compared with FD_{BC} [13] and SFD [10] methods when applied to a clinically adjusted subsample of BMES population database. The results show that FD-C had a significant relationship with the 10-year indicator of CHD and stroke and thus was a significantly better predictor of CHD and stroke ($p = 0.016$, $\alpha = 0.05$) while other fractal measures did not show any significant association for these subsamples (all p-values > 0.05). The median FDC for the entire image was lower (1.981) for the cases compared to the controls (1.987). This indicates that reduction in the complexity of the retinal vasculature is an indicator of disease, and is comparable with findings of Lipsitz et al. [14] which is associated with functional loss [15], Such loss of complexity has also been observed to be associated with ageing and disease in cardiac activity [16], neural system [17], electromyogram (EMG) [18–20] and general physiological measures [15].

12.3 Diabetes and Retinal Fractals

As discussed in Chapter 8, Section 8.6, changes in the complexity of retinal vessels can be associated with different stages of diabetes and its complications such as diabetic retinopathy (DR). However, what happens before the onset of diabetic retinopathy? The aim of this case was to statistically model the association between retinal vascular parameters with type II diabetes where there was no reported retinopathy.

The study was conducted using retinal images from the Indian population with mild-non proliferative and minimal diabetic retinopathy. For this purpose, FD of retinal vasculatures together with two new vascular parameters, including total number and average variability of the acute branching angles were used for early detection of diabetes and the underlying micro vascular complications.

12.3.1 Materials

Experiments were conducted using database collected in Department of the Retina, Save Sight Centre hospital located in Delhi. Approval for this study was granted by the Human Research Ethics Committee (HREC) of the Royal Melbourne Institute of Technology (RMIT University), Melbourne, Australia and also by the institutional review board at Save Sight Centre hospital in accordance with the declaration of Helsinki (1975, as revised in 2004). All participants were respondents to a request advertised in the 'Save Sight Centre' in Delhi. The purpose and experimental procedure in plain language was given to the participants in writing and also explained verbally. Written and oral consent was obtained from the participant prior to data collection.

All the volunteers self-evaluated themselves to be 'reasonably active' and none of them were pregnant. The participants were classified in two groups of type-II diabetes (case) with no observable retinopathy and non-diabetic (control) patients. The diabetic cases were confirmed by the patients' physician based on either (i) their fasting or (ii) post-prandial glucose plasma levels being greater than 126 mg and 200 mg/decilitre respectively. None of the diabetic patients had any observable intra-retinal haemorrhages or venous beading, hard exudates and neovascularisation according to the classification levels by International Clinical Disease Severity Scale for DR. This was confirmed by an ophthalmologist after examination of the both eyes.

The participants' demographic information including age, gender, weight (kg), height (m), systolic and diastolic blood pressure (mmHg), skin fold (mm), cholesterol level (LDL & HDL), were recorded. The participants' age was limited to a narrow range to remove confounding effect of the age factor on the analysis outcome; and provide a better balance between the number of diabetic cases and the control groups. All the participants were non-smoker, did not consume alcohol, had no history of any cardiovascular disease and did not have any history of antihypertensive and lipid-lowering medications.

One Optic Disk Centred (ODC) and one Macula Centred (MC) fundus image was taken from both eyes of each subject making total number of four images for each participant. The photographs were taken in mydriatic mode, in a dimmed light room, using a mydriatic Kowa Vx alpha camera

(Kowa, Japan). The original image resolution was 300dpi (4288 × 2848 pixels) and the camera field of view was set to 30°.

12.3.2 Method

All the images were examined in pairs for quality assurance (i.e., vessel to background contrast and illumination artefacts). After quality assurance and discarding upgradable images, a total number of 180 retinal images (Two (Left & Right) × Two (ODC & MC) × 45 subjects (13 diabetics and 32 non-diabetics)) were obtained. All the original images were cropped and re-sampled to identical sizes of 729 × 485 pixels. Image enhancement and segmentation (bineraziation) was performed to reduce the background artifacts and improve vessel to background contrast.

Retinal vascular geometry features were measured automatically using Retina Vasculature Assessment Software (RIVAS), an established software package based in MATLAB®, MathWorks, Inc, USA; developed by the authors. In brief, the software combines the individual measurement into summary indices of multiple measurement options. Some of these are; (i) vessel calibre of a specified segment, (ii) simple tortuosity, (iii) number of different fractal dimension (FD), (iv) vessel to background ratio/percentage (V/B (%)) and (v) average of the acute branching angles (ABA) defined as the smallest angle between two daughter vessels and (vi) the total number of branching angles (TBA). The FD measures include Binary Box-Counting (FD_{BC}) and Differential Box-Counting (FD_{DBC}) [21] and Fourier (Spectrum) Fractal Dimension (FFD) as explained in Chapter 7.

In this study, simple tortuosity was measured as the ratio between the actual length of a vessel segment and the shortest (Euclidean) distance between the two endpoints within the same segment providing a reflection of the shape/curvature of the vessel. FD_{BC} was calculated on skeletonized images as indicator of vascular network complexity without comprising any vessel calibre information.

Vessel diameter summary was also measured using IVAN software (University of Wisconsin, Madison, WI, USA) based on the calibre summary of the biggest 6 arterioles and venules separately, represented by Central Retinal Arteriolar Equivalent (CRAE) and Central Retinal Venular Equivalent (CRVE), [22] as well as the ratio of the calibre of arterioles to venules (AVR). The measurements were performed within a fixed region between 0.5 to 1.0 optic disc diameter from the disc margin. CRAE and CRVE were obtained based on the revised Knudtson-Parr-Hubbard formula [22].

In this study, ODC and MC images were analysed separately but for each image category (i.e., ODC/MC) the retinal vascular parameters of the left and right eyes of each subject were averaged prior to analysis.

12.3.3 Data analysis

Statistical analyses were performed using Minitab® v.16.1.0 and R studio (R® statistical software v.3.3.0). As the number of observations was relatively small, the data was statistically up-sampled to the new size of 200 samples (i.e., 58 diabetic cases and 142 non-diabetic) using the bootstrapping technique with sample replacement. The up-sampled data was then standardized and centred to decrease the multi-co-linearity between an interaction term and its corresponding main effects as well as making categorical parameters such as gender, comparable with continuous parameters.

Sixteen predictors were used in this analysis; vessel tortuosity (both mean and standard deviation (SD)), ABA , SD of angle, TBA, CRAE, CRVE, AVR, FD_{BC} and VB plus the patient's demographic information (i.e., Gender, age, systolic and diastolic blood pressure, Body Mass Index (BMI), and skin fold). A linear regression model was built using all the 16 parameters as dependant variable and diabetes status with two possible categories (i.e., case and control) as independent variable. Analysis of variance (ANOVA) test was performed and F statistic was calculated to check the model fit. R-squared was also calculated to examine whether the model is close to the regression line and obtain the percentage from the dependant variable's variation explained by the model. For each predictor in the model the coefficients and their significance level were calculated to identify potential non-significant variables and remove them from the model. The test for multi-co-linearity was performed by looking into the Variance Inflation Factor (VIF) for standard error of the regression coefficients. VIF greater than 5 was considered as presence of high multi-co-linearity between the predictors.

This study has taken the stepwise regression approach, where the aim was to improve the exploratory stages of model building, but without compromising the physiological understanding of the predictors by using methods such as principal component analysis (PCA) for dimensionality reduction. This is essential for medical applications because the clinicians are keen to identify the relevant health parameters along with improved labelling of the data. For this purpose, stepwise regression analysis was performed to select the variables that are significantly important. In this process the most important variables are first selected with a forward searching algorithm followed by a backward elimination process to provide a reduced model with most suitable variables. This method adds and removes the predictors in each step until all the variables used in the model have $p\_value <= \alpha$ and the one which are not used in the model have $p\_value > \alpha$ with $\alpha = 0.05$. For each predictor in the reduced model, VIF was calculated to test for the multi-co-linearity followed by testing the model for fitting performance using ANOVA and R-squared statistics.

12.3.4 Results

The participants' demographic information (mean±SD) at the baseline, prior to bootstrapping the data, contained gender—Female/Male (Diabetic = 7/6, Non-diabetic = 12/20), age (Diabetic = 56 ± 7.25 years, Non-diabetic = 51.93 ± 9.0 years), BMI (Diabetic = 27.34 ± 3.97, Non-diabetic = 26.52 ± 6.42) defined as weight (kg) divided by squared height (m^2), systolic blood pressure (Diabetic = 143.84 ± 21.90 mmHg, Non-diabetic = 131.55 ± 12.53 mmHg) and diastolic blood pressure (Diabetic = 81.15 ± 11.20 mmHg, Non-diabetic = 80.34 ± 11.48 mmHg) and skin fold (Diabetic = 38.69 ± 4.11 mmHg, Non-diabetic = 36.86 ± 6.02 mmHg).

Data analysis was performed separately on both ODC and MC images to explain the relationship between potential predictors and diabetes factor. The first explanatory model was built for the ODC images. The result from linear regression analysis and ANOVA test for the first full model is shown in Table 12.3.

Table 12.3. Predictors coefficients, significance level for the first model and ANOVA test result.

Predictors	Coefs.	SE	T	*P*	VIF	Adjusted R^2	ANOVA	
							F	*P*
Constant	0.46335	0.01242	37.31	< 0.001		96.6%	3094.49	< 0.001
CRAE	11.9878	0.0931	128.81	< 0.001	1493.583			
CRVE	–4.20147	0.03378	–124.39	< 0.001	238.956			
AVR	–11.7099	0.0905	–129.43	< 0.001	1631.594			
Mean Tortuosity	–0.5174	0.01389	–37.26	< 0.001	55.644			
SD of Tortuosity	0.150746	0.008002	18.84	< 0.001	18.545			
ABA	2.54491	0.02117	120.2	< 0.001	49.765			
SD of Angle	1.68918	0.01853	91.14	< 0.001	36.814			
VB	0.82474	0.02233	36.93	< 0.001	155.256			
TBA	0.960501	0.009531	100.77	< 0.001	22.108			
FD	–0.81633	0.01878	–43.47	< 0.001	94.546			
Gender	1.1179	0.0097	115.27	< 0.001	18.8			
Age	–0.53138	0.01118	–47.52	< 0.001	12.804			
BMI	0.02697	0.011	2.45	0.015	13.547			
Systolic blood pressure	0.441949	0.004433	99.71	< 0.001	5.542			
Diastolic blood pressure	–0.22795	0.005911	–38.56	< 0.001	4.476			
Skin fold	–0.32616	0.01465	–22.27	< 0.001	19.118			

In this model, the adjusted R-squared of 96.6% indicates 96.6% from the diagnosis variation is due to the model (or due to change in predictors) and only 3.4% is due to error or some unexplained factors. *P* values of < 0.05 represent the predictors are related to diabetes at α level of 0.05. The ANOVA test also shows that the linear regression model fits well to the data (F = 3094.49, $P < 0.001$). However, in this model VIF is greater than 5 for almost all the predictors (except diastolic blood pressure) showing that there is problem concerning the predictors' co-linearity. Therefore, a second model was built from the full model using stepwise procedure to reduce the number of parameters and the multi-co-linearity between the predictors. The result has been provided in Table 12.4.

Table 12.4. Predictors coefficients, significance level for the second model and ANOVA test result.

Predictors	Coefs.	SE	T	*P*	VIF	Adjusted R^2	ANOVA	
							F	*P*
Constant	0.312	0.015	20.11	< 0.001				
CRAE	–0.088	0.018	–4.71	< 0.001	1.47			
Mean Tortuosity	0.219	0.027	8.13	< 0.001	3.01			
ABA	0.239	0.038	6.24	< 0.001	4.1			
SD of Angle	0.179	0.040	4.45	< 0.001	4.78	44.95%	50.11	< 0.001
VB	–0.269	0.030	– 8.88	< 0.001	3.82			
TBA	0.262	0.022	11.48	< 0.001	2.17			
Age	0.111	0.020	5.44	< 0.001	1.73			
Systolic blood pressure	0.108	0.017	6.11	< 0.001	1.31			

Hence, the final explanatory model for diabetes can be represented by the following equation:

$$Diabetes = -0.088\ CRAE + 0.219\ Mean\ Tortuosity + 0.239\ ABA + 0.179\ SD\ of\ Angle - 0.269\ VB + 0.262\ TBA + 0.11\ Age + 0.108\ Systolic\ Blood\ Pressure + 0.312$$

For this reduced model, ANOVA test shows the relationship between diabetes and the predictors is significant at α level of 0.05 (F = 50.11, $P < 0.001$). The VIF factor in all cases is smaller than 5 showing negligible co-linearity between predictors. In this model the coefficients are weaker than the first model with reduced R-squared. This result is in line with general expectations that there is reduction in model goodness of fit with reduction in number of features; however, this does not represent decline in the explanatory power of the reduced model compared to the full model. Also interpretation of the coefficients in the second model with reduced number of variables and negligible degree of co-linearity is more valid and accurate compared to the full model. Comparison between the two models shows that, some coefficients have lower magnitude in the second model. Also some predictors (i.e., CRAE, Mean Tortuosity, VB and age) change their sign from the full model to the reduced model, resulting into some degree of uncertainty for the interpretation of the full model with highly co-linear predictors.

The results also shows that the retinal vasculature parameters with α level of 0.05 that play a significant role in the reduced explanatory model of diabetes are the CRAE, Mean Tortuosity, ABA, SD of branch angles, VB, and TBA. From the clinical and demographical information, only systolic blood pressure and age were found to be significant predictors. The same analysis, as explained above, was performed on MC images; however, no model was found to provide adequate fit to the data for this database.

12.3.5 Discussion and conclusion

This research has proposed an explanatory model for the association between retinal vasculature parameters and diabetes in the Indian population with no DR. The analyses were performed on both, MC and ODC images using a 30° non-mydratic eye-fundus camera. The significance of this work is that it reports automatic analysis of the eye-fundus images and provides an explanatory model for early changes in some retinal vascular parameters as a result of diabetes. This study has also introduced two new retinal vascular parameters, including TBA and ABA; and employed them together with other predictors and patients demographic information to create an explanatory model for prediction of diabetes in the absence of DR.

In this work, linear regression analysis was performed to model the association between a large number of explanatory variables as the predictors (i.e., retinal vascular parameters and clinical information) and DM as the response variable. The application of stepwise regression allowed for dimensionality reduction as well as solving the multi-co-linearity problem, while making the send model clinically interpretable. It was important that predictors should be clinically relevant and without compromising the physiological understanding of the predictors. The result showed that six retina vascular parameter of (1) CRAE, (2) Mean tortuosity, and (3) ABA, (4) SD of angle (5) VB and (6) TBA were associated with diabetes in Indian population when there was no observable DR.

The significant outcome of this study is that it provides the basis for an alternate technique to detect diabetes among people with no DR. This could be very useful for people who are hesitant in taking blood tests.

Another major outcome of this study is that it has introduced two retinal vascular parameters may find application in a predictive model to predict the risk of developing DR in diabetic patients with no DR. Therefore, it is proposed to conduct longitudinal study to monitor the progress of the patients and identify changes among those who may later develop DR. The measurements are suitable for automation and for being used with simple eye-fundus imaging.

However, this study was limited in terms of the population type, having been conducted only in Delhi in India and represented narrow demographics with limited number of samples. This may lead to limited diagnostic usability for the current model. It is essential to include subjects with similar cultural, ethnic, and socio-economic conditions. There is also the need for conducting similar tests on larger demographic data to observe potential differences and better evaluate the performance of the proposed model.

12.4 Muscle Fatigue and Fractal Properties

Localized muscle fatigue is the state when the ability of a skeletal muscle to contract or produce force is highly diminished and this condition is local to a set of muscles when the neuro-stimulation pathways are intact [23]. This generally occurs after sustained or intense contractions. Numerous researchers have studied the surface electromyogram (sEMG) signal to obtain a non-invasive and objective measure of muscle fatigue. Researchers have considered various signal features such as root mean square (RMS), median frequency [24,25] wavelet transforms [26,27], fractal dimension [28], and normalised spectral moments [29,30] to identify fatigue.

Changes in motor unit recruitment patterns are associated with muscle fatigue, with increased synchronization associated with the onset of localized muscle fatigue [31]. This may occur due to increased central drive leading to synaptic input that is common to more than one neuron, or reduction in conduction velocity or a combination of both. Kleine et al. [31] posits synchronization must be responsible for the spectral shift to lower frequencies not attributable to a conduction velocity change.

Kumar et al. [32] reported that fatigue will increase synchronization of motor units and lead to increased dependence between the activities recorded from different sections of the muscle. Fractal dimension (FD) of sEMG has been used to study and characterize levels of muscle activation as explained in Chapter 5.

In this section, we are reporting a case study using data and materials, the following has been published as a journal article [32].

12.4.1 Materials

Surface Electromyogram (sEMG) signals were recorded using Delsys (Boston, MA, USA), a proprietary sEMG acquisition system. The system supports bipolar recording and has a fixed gain of 1000, CMRR of 92 dB and bandwidth of 20–450 Hz, with 12 dB/octave roll-off. The sampling rate is fixed at 1000 samples/second, and the resolution is 16 bits/sample. Two proprietary bipolar electrodes of Delsys (Boston, MA, USA) were placed on the skin of the participant's overlying the muscle under investigation. The active electrodes were dry, bar-shaped (1 mm wide and 10 mm long) silver electrodes. These electrodes had the two electrodes (bars) mounted directly on the preamplifier with fixed inter-electrode distance being 10 mm.

The electrodes were placed on the anterior of the arm above the biceps. The distance between the two channels was maintained at 2 cm. A reference electrode was placed on the dorsal section and under the elbow. Prior to electrode placement, the skin area was cleaned with alcohol swabs and exfoliated to reduce skin impedance and ensure good adhesion of the electrodes.

Twenty five subjects (20 male, 5 female aged 25–30) volunteered to participate in these trials. The experiments were approved by the RMIT University Human Ethics Committee. During the experiments, the volunteers were seated in a sturdy and adjustable chair with their feet flat on the floor and their upper arm was rested on the surface of an adjustable desk such that the forearm was vertical. The elbow was fixed at 90 degrees, with the fingers in line with a wall mounted force sensor (S type force sensor (Interface SM25)) attached to a comfortable hand sized ring with a flexible steel wire, and the ring was held in the palm of the hand. To determine the maximal voluntary contraction (MVC), three maximal contractions of 5 seconds were performed with 120 seconds rest time between each effort. The participants pulled the ring and the force of contraction was recorded. The average of these readings was considered to be the MVC. If there were any outliers, the experiment was repeated.

In the first set of experiments, the participants performed isometric contractions at 25%, 50%, 75% and 100% MVC. Participants were asked to perform contractions until they felt exhausted. The duration of the contraction was referred to as the endurance period, Te, and this was found to be different for different participants and for different levels of muscle contraction. It was based on 0 to 10 pain index scale with 0 corresponding to 'No pain', and 10 corresponding to 'Maximum pain'. A score of 8 and above corresponded to muscle fatigue. Between each contraction, the participants were given a rest period which was minimum of 15 minutes, but as long as they required for the pain level to become less than 4.

12.4.2 Methods

In order to understand the change in the fractal properties of the muscle fatigue, fractal dimension of the EMG data recorded during fatiguing contraction was computed. For this case study, the EMG signal recorded during 75% of the MVC contraction was considered. Fractal dimension of the sEMG was computed using Higuchi algorithm as explained in Chapter 3, Section 3.3.

The data was segmented to identify 5 equidistant time locations over the duration of the exercise. The start of the exercise was labelled as To and the end of the endurance of the participant for that specific contraction exercise was labelled as Te. 0.25Te, 0.5Te and 0.75Te corresponded to 0.25, 0.5 and 0.75 of the Te. Fractal Dimension of surface EMG was computed using a moving one second (1000 samples) window.

12.4.3 Results and discussion

Figure 12.2 shows the mean and standard deviation of the FD of sEMG of all healthy young subjects recorded at their 75% MVC. From the figure it is observed that there is a gradual decrease in Fractal dimension over the duration of the contraction and a large inter-subject variation. This change in the fractal properties can be attributable to the changes due to MU synchronization during muscle fatigue as reported by Mesin et al. [33]. They have also reported that decrease in FD can be an indicator of the progressive MU synchronisation.

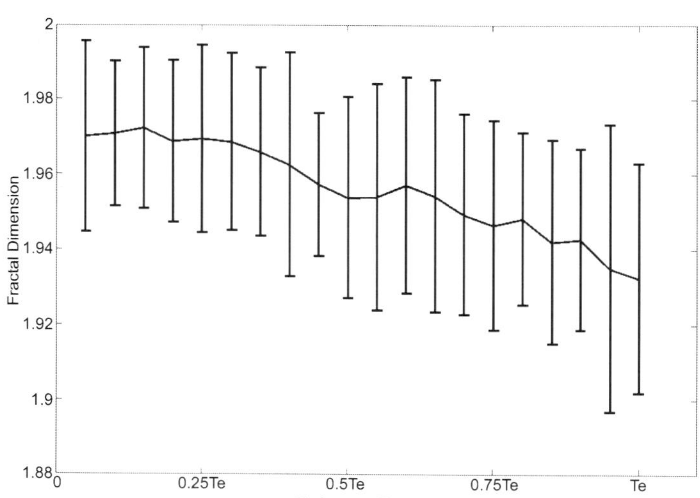

Figure 12.2. Mean (SD) fractal dimension of EMG as a progression of endurance time.

Studies by Holtermann et al. [34] have reported the changes in MU synchronization with fatigue using sub-band skewness of sEMG based on the sampling of the large motor unit samples during the initial and final values phase of the recording. Beretta-Piccoli et al. [35] reported that FD is not significantly affected by changes in the muscle conduction velocity, and is most related to the MU synchronization. The results show that there is a change in the FD of sEMG when there is a change in the muscle status due to the muscle fatigue.

12.5 Summary

This book has shown the fractal properties of biomedical signals and images. This chapter has shown the association of the FD of these with disease conditions. For this purpose, three examples have been given where the fractal analysis of biomedical signals and images has been measured and the difference between the values of case and control has been investigated.

This chapter has demonstrated possible applications of fractal geometry based analysis of biomedical signals and images. One significant advantage of using fractal dimension as a feature to associate with disease is that it is a global to the image or the signal, and often can be performed without the need for segmentation. This makes it very suitable for automated machine based assessment for disease and other conditions. Risks of stroke, early stages of diabetes and muscle fatigue are major issues in our society and methods to assess or monitor these can be very significant.

References

1. Aliahmad, B., D.K. Kumar, H. Hao, P. Unnikrishnan, M.Z. Che Azemin, R. Kawasaki and P. Mitchell. 2014. Zone specific fractal dimension of retinal images as predictor of stroke incidence. The Scientific World Journal, 2014, p. 7.
2. Mitchell, P., W. Smith, K. Attebo and J.J. Wang. 1995. Prevalence of agerelated maculopathy in Australia. The Blue Mountains Eye Study. Ophthalmology, 102: 1450–60.
3. Mitchell, P., J.J. Wang, T.Y. Wong, W. Smith, R. Klein and S.R. Leeder. 2005. Retinal microvascular signs and risk of stroke and stroke mortality. Neurology, 65: 100–9.
4. Che Azemin, M.Z., D.K. Kumar, T.Y. Wong, J.J. Wang, R. Kawasaki and P. Mitchell. 2010. Retinal stroke prediction using logistic-based fusion of multiscale fractal analysis, in Imaging Systems and Techniques (IST), 2010 IEEE International Conference on, pp. 125–128.
5. Azemin, M.Z.C., D.K. Kumar, T.Y. Wong, R. Kawasaki, P. Mitchell and J.J. Wang. 2011. Robust methodology for fractal analysis of the retinal vasculature. Medical Imaging, IEEE Transactions on, 30: 243–250.
6. Soares, J.V.B. 2008. HM integration for vessel segmentation. Available: http://sourceforge.net/projects/retinal/files/mlvessel/.
7. Higuchi, T. 1988. Approach to an irregular time series on the basis of the fractal theory. Physica D: Nonlinear Phenomena, 31: 277–283.
8. Ahammer, H. 2011. Higuchi Dimension of Digital Images. PLoS ONE, 6: e24796.

9. Cheung, N., K.C. Donaghue, G. Liew, S.L. Rogers, J.J. Wang, S.W. Lim, A.J. Jenkins, W. Hsu, M.L. Lee and T.Y. Wong. 2009. Quantitative assessment of early diabetic retinopathy using fractal analysis. Diabetes Care, 32: 106–10.
10. Kawasaki, R., M.Z. Che Azemin, D.K. Kumar, A.G. Tan, G. Liew, T.Y. Wong, P. Mitchell and J.J. Wang. 2011. Fractal dimension of the retinal vasculature and risk of stroke: A nested case-control study. Neurology, 76: 1766–1767, May 17, 2011.
11. Azemin, M.Z.C., D.K. Kumar, T.Y. Wong, J.J. Wang, P. Mitchell, R. Kawasaki and H. Wu. 2012. Age-related rarefaction in the fractal dimension of retinal vessel. Neurobiology of Aging, 33: 194.e1–194.e4.
12. de Mendonca, M.B., C.A. de Amorim Garcia II, R. de Albuquerque Nogueira III, M.A.F. Gomes, M.M. Valença and F. Oréfice. 2007. Fractal analysis of retinal vascular tree: segmentation and estimation methods. Arq. Bras. Oftalmol., 70: 413–22.
13. Masters, B.R. 2004. Fractal analysis of the vascular tree in the human retina. Annu. Rev. Biomed. Eng., 6: 427–52.
14. Lipsitz, L.A. and A.L. Goldberger. 1992. Loss of complexity and aging: Potential applications of fractals and chaos theory to senescence. JAMA, 267: 1806–1809.
15. Kyriazis, M. 2003. Practical applications of chaos theory to the modulation of human ageing: nature prefers chaos to regularity. Biogerontology, 4: 75–90.
16. Pikkujamsa, S.M., T.H. Mäkikallio, L.B. Sourande, I.J. Räihä, P. Puukka, J. Skyttä, C.-K. Peng, A.L. Goldberger and H.V. Huikiri. 1999. Cardiac interbeat interval dynamics from childhood to senescence: comparison of conventional and new measures based on fractals and chaos theory. Circulation, 100: 393.
17. Schierwagen, A. 1987. Dendritic branching patterns. Chaos in Biological System, pp. 191–193.
18. Kaplan, D.T., M.I. Furman, S.M. Pincus, S.M. Ryan, L.A. Lipsitz and A.L. Goldberg. 1991. Aging and the complexity of cardiovascular dynamics. Biophys. J., 59: 945–9.
19. Kresh, J.Y. and I. Izrailtyan. 1998. Evolution in functional complexity of heart rate dynamics: a measure of cardiac allograft adaptability. American Journal of Physiology-Regulatory, Integrative and Comparative Physiology, 275: 720.
20. Skinner, J.E. 1994. Low-dimensional chaos in biological systems. Nature Biotechnology, 12: 596–600.
21. Sarkar, N. and B.B. Chaudhuri. 1994. An efficient differential box-counting approach to compute fractal dimension of image. Systems, Man and Cybernetics, IEEE Transactions on, 24: 115–120.
22. Che Azemin, M.Z., D.K. Kumar, T.Y. Wong, R. Kawasaki, P. Mitchell and J.J. Wang. 2011. Robust methodology for fractal analysis of the retinal vasculature. Medical Imaging, IEEE Transactions on, 30: 243–250.
23. Hagg, G.M. 1992. Interpretation of EMG spectral alterations and alteration indexes at sustained contraction. J. App. Physiol., 73: 1211–1217.
24. Merletti, R. and S. Roy. 1996. Myoelectric and mechanical manifestations of muscle fatigue in voluntary contractions. J. Orthop. Sports Phys. Ther., 24(6): 342–53.
25. Merletti, R. and P. Parker. 2004. Electromyography, New York John Wiley and Sons.
26. Kumar, D.K., N.D. Pah and A. Bradley. 2003. Wavelet analysis of surface electromyography. IEEE Transactions on Neural Systems and Rehabilitation Engineering, 11(4): 400–406.
27. Dimitrios, M., I. Hostens, G. Papaioannou and H. Ramon. 2005. Dynamic muscle fatigue detection using self-organizing maps. Applied Soft Computing, 5(4): 391–398.
28. Wang, G., X.M. Reng, L. Li and Z.Z. Wang. 2007. Multifractal analysis of surface EMG signals for assessing muscle fatigue during static contraction. Journal of Zhejiang University – Science (English translations through Springer), 8: 910–915.
29. Dimitrov, G.V., T.I. Arabadzhiev, K.N. Mileva, J.L. Bowtell, N. Crichton and N.A. Dimitrova. 2006. Muscle fatigue during dynamic contractions assessed by new spectral indices. Medicine & Science in Sports & Exercise, 38: 1971–1979.
30. Dimitrova, N.A., T.I. Arabadzhiev, J.Y. Hogrel and G.V. Dimitrov. 2009. Fatigue analysis of interference EMG signals obtained from biceps brachii during isometric voluntary

contraction at various force levels. Journal of Electromyography and Kinesiology, 19(2): 252–258.
31. Kleine, B.U., D.F. Stegeman, D. Mund and C. Anders. 2001. Influence of Moto neuron firing synchronization on SEMG characteristics in dependence of electrode position. J. Appl. Physiol., 91: 1588–1599.
32. Kumar, D.K., S.P. Arjunan and G.R. Naik. 2011. Measuring increase in synchronization to identify muscle endurance limit, in IEEE Transactions on Neural Systems and Rehabilitation Engineering, 19(5): 578–587.
33. Mesin, L., C. Cescon, M. Gazzoni, R. Merletti and A. Rainoldi. 2009. A bi-dimensional index for the selective assessment of myoelectric manifestations of peripheral and central muscle fatigue. J Electromyogr Kinesiol, 19: 851–863.
34. Holtermann, A., C. Gronlind, J.S. Karlsson and K. Roeleveld. 2009. Motor unit synchronization with fatigue: Described with a novel sEMG method based on large motor unit samples. Journal of Electromyography and Kinesiology, 19: 232–241.
35. Beretta-Piccoli, M., G. D'Antona, M. Barbero, B. Fisher, C.M. Dieli-Conwright, R. Clijsen and C. Cescon. 2015. Evaluation of Central and Peripheral Fatigue in the Quadriceps Using Fractal Dimension and Conduction Velocity in Young Females. PLoS ONE, 10(4): e0123921.

Index